Cerebral Dominance

Cerebral Dominance

The Biological Foundations

edited by Norman Geschwind
and Albert M. Galaburda

Harvard University Press
Cambridge, Massachusetts
and London, England

Copyright © 1984 by the President and Fellows of
  Harvard College
All rights reserved
Printed in the United States of America
10 9 8 7 6 5 4 3

Library of Congress Cataloging in Publication Data
Main entry under title:
Cerebral dominance.
  Bibliography: p.
  Includes index.
  1. Cerebral dominance. I. Geschwind, Norman,
1926–   . II. Galaburda, Albert M., 1948–
[DNLM: 1. Dominance, Cerebral — Congresses.
2. Dominance, Cerebral — Physiology — Congresses.
3. Brain mapping — Congresses.   WL 355 C414 1983]
QP385.5.C46   1984   612'.825   84-6574
ISBN 0-674-10658-X (cloth)
ISBN 0-674-10659-8 (paper)

To Jane and Peter Pattison
without whose concern for the dissemination
of new knowledge about the role of the brain
in behavior this volume might not have
come into being

Preface

The growth of modern science is often thought to reflect the development of techniques that make it possible to study phenomena concealed from the senses. The microscope, telescope, X-ray, and recording electrode have revealed a hidden world and led many to believe that all the remaining mysteries must lie beyond the capacities of ordinary observation. Yet, as Goethe pointed out, what lies directly in front of us may fail to be noticed —and for that very reason may fail to be investigated.

Cerebral dominance is a dramatic example of this principle. Throughout history its most obvious manifestations, the preferential use of the right hand by the majority of humankind and the existence of a nondextral minority, have been universally recognized. In every society primitive beliefs have arisen about the origins and associations of these conditions, beliefs reflected in our everyday vocabularies in words and phrases such as "sinister," "gauche," "left-handed compliment," and many others.

One might expect that these remarkable facts would have tantalized scientists and generated efforts to comprehend one of the cardinal biological features of humans. Despite the dramatic discovery over a century ago that in most individuals the ability to acquire language is predominantly dependent on the left hemisphere, early scattered attempts to penetrate to the foundations of this phenomenon were generally neglected. It is intriguing to speculate on why this is so. And it is equally intriguing to speculate on why, less than two decades ago, there arose a group of investigators, working independently in very different disciplines, who began to explore intensively the biological foundations and associations of cerebral dominance.

Although the total number of researchers is still very small, the rapid growth of this new area stimulated us to organize a conference devoted entirely to biological studies of dominance over the previous fifteen years. This meeting, held in Boston on April 4-6, 1983, brought together a group of investigators each of whom is a pioneer in some aspect of the new discipline. This volume is an outgrowth of that conference. The obvious enthusiasm of both the listeners and the participants made it clear that what could be heard by only a few should be made available to a much wider audience.

The conference could not have taken place without the generous assistance of Jane and Peter Pattison of the Institute for Child Development Research. We are deeply indebted to them for their support then, and subsequently in the production of this book.

We are grateful also to Loraine Karol for her invaluable help in organizing the meeting and preparing the volume, and to Susan Wallace and Vivian Wheeler of Harvard University Press.

N.G.
A.M.G.

Contents

1 Historical Introduction   1
Norman Geschwind

Part One   Brain Asymmetry in Humans

2 Anatomical Asymmetries   11
Albert M. Galaburda

3 Radiological, Developmental, and Fossil Asymmetries   26
Marjorie LeMay

4 A Dendritic Correlate of Human Speech   43
Arnold B. Scheibel

5 Brain Electrical Activity Mapping   53
Frank H. Duffy, Gloria B. McAnulty, and Steven C. Schachter

6 Asymmetrical Lesions in Dyslexia   75
Thomas L. Kemper

Part Two   Brain Asymmetry in Other Species

7 Learning, Forgetting, and Brain Repair   93
Fernando Nottebohm

8 Behavioral Asymmetry   114
Victor H. Denenberg

9 Age, Sex, and Environmental Influences   134
Marian Cleeves Diamond

10 Functional and Neurochemical Asymmetries   147
Stanley D. Glick and Raymond M. Shapiro

11	Lateralization of Neuroendocrine Control *Ida Gerendai*	167
12	Experimental Modification of Gyral Patterns *Patricia S. Goldman-Rakic and Pasko Rakic*	179

Part Three Biological Associations of Laterality

13	Twinning, Handedness, and the Biology of Symmetry *Charles E. Boklage*	195
14	Laterality, Hormones, and Immunity *Norman Geschwind and Peter O. Behan*	211

Contributors 225

Index 227

Cerebral Dominance

Chapter 1

Historical Introduction

Norman Geschwind

Although cerebral dominance has been studied for more than 120 years, its biological foundations and relationships have received only scant attention. Within the past 15 years, however, a few investigators have turned their attention to this previously neglected area. This volume is devoted entirely to the results of these new approaches, which have quickly broken down the barriers that for so long separated research on dominance from the general field of biology. The studies reported here are only the very first steps in delineating the mechanisms and associations of laterality. Yet on the basis of the contributions in these pages we can confidently predict that within the next 25 years many kinds of biological techniques will be devoted to elucidating the processes by which the two sides of the nervous system attain different specializations. Furthermore, we are becoming aware that lateralization of the nervous system plays a previously unexpected role in many aspects of structure and function. The study of dominance is thus likely to contribute to many areas of biology and medicine to which it was previously considered irrelevant.

The new developments can probably be best understood if they are viewed in the context of the entire history of the field of cerebral dominance, of which even a brief survey reveals that long-held beliefs have repeatedly turned out to be incorrect. The biological approaches reviewed in this volume have led to similar rejections.

It is sometimes stated that Marc Dax first described left-hemisphere dominance for language in 1836, but this assertion is controversial. In any case, it was not until the observations of Paul Broca in the 1860s that the concept of cerebral dominance entered the world of scientific thinking. Broca reported that aphasia (disturbance of language function resulting from brain lesions) was the result of destruction of delimited regions on one side of the brain. Broca found in his original series of cases that the aphasia-producing lesion was always situated on the left side. The surprising finding that a complex learned ability depended on structures on one side of the brain led to formulation of the concept of cerebral dominance, a term which means that one hemisphere is exclusively or preponderantly responsible for the acquisition and/or performance of a given function.

The fact that the lesions in all of his cases lay on the left side led Broca to postulate that there was a close relationship between dominance for language and the recognized preponderance of right-handedness in human populations. Right-handers were thought to have left-hemisphere language dominance, and the reverse was thought to be true in left-handers. Although the tenet that dominance for handedness and language are intimately linked has been shown repeatedly to have serious shortcomings, it has continued to exert a major, and in my view unfortunate, influence.

Another major concept that developed in the last half of the nineteenth century was that one hemisphere, usually the left, was dominant, while the other was subordinate. It was commonly accepted that the left hemisphere was the storehouse of all complex activities and that the right, nondominant hemisphere played only a minor role. Despite occasional evidence that the right hemisphere might have certain special functions, this idea held sway for a very long period. It runs counter, of course, to the basic views of most biologists, who would argue that it is unlikely that a huge volume of metabolically expensive nervous tissue would be maintained for so long in the course of evolution if so much of it possessed no important function.

These two concepts continued to dominate thinking until after the Second World War. As active investigators in several countries found that lesions of the right hemisphere produced distinctive disorders, they realized that it was not possible to speak of a dominant and a subordinate hemisphere; each had its own strong specializations. There are still many individuals who, despite the repeated finding of complementary specialization in the two hemispheres, subscribe to the view that the right hemisphere never achieves the high intellectual level conferred on the left by its possession of the capacity for language. This belief is reflected in the common but fallacious notion that the left hemisphere is not only verbal but logical. As evidence accumulates, however, it appears likely that this last bastion of left-hemisphere superiority will have to fall and the hemispheres will be regarded as equally advanced in the complexity of their contrasting functions.

The original view of Broca that in the right-handed majority of the population language lay on the left, whereas in the left-handed minority language was located on the right, also had to be abandoned. It had long been known that "crossed" aphasias existed, that is, that a small minority of right-handers became aphasic after right-hemisphere lesions and that in some sinistrals left-sided lesions had the same effect. The major fact that overthrew the traditional concept was the discovery that even in left-handers a distinct majority of aphasia-producing lesions lay on the left side. This finding led in turn to a new concept, which held sway for a time; but it too is now under fire. It was argued that in a majority (perhaps 60%) of left-handers language dominance lay in the left hemisphere, while in 40% it was in the right hemisphere. On the basis of studies dating back to the late 1950s, another possibility has been advanced: that in left-handers, brain dominance is less marked. It has been argued that left-handers are more likely to become aphasic with a lesion on either side, but that they have a higher rate of recovery.

Certain views concerning dominance were even more persistent than the belief in a single dominant hemisphere and in the close relationship of handedness and language laterality. Although an anatomical basis for dominance had been suggested by several investigators, it was generally accepted well into the 1960s that there was no significant anatomical asymmetry that could account for the striking differences in function of the two hemispheres. Toward the end of that decade, however, Norman Geschwind and Walter Levitsky (1968) reported a large gross asymmetry in the cortex of the posterior superior temporal region—in other words, in part of the temporal language area of Wernicke. Albert Galaburda and associates (1978) went on to report that this asymmetry in favor of the left side at the gross anatomical level reflected the greater size of a region of particular cytoarchitectonic structure.

Another of the long-held basic postulates of the study of dominance was that it was an exclusively human characteristic. Fernando Nottebohm (1977), however, overturned this view by his demonstration of unilateral dominance for song in several species of birds. Victor Denenberg (1981) has shown right unilateral dominance for spatial function and emotion, a pattern similar to that of humans, in the rat.

These results were rapidly succeeded by other pioneering observations. Marjorie LeMay and her coworkers (1972, 1976, 1978) demonstrated convincing anatomical asymmetry in the living human brain by radiological methods. LeMay and Geschwind (1975) showed by anatomical studies that the asymmetries of the sylvian fissures seen in humans were also found in the brains of the great apes, but not in monkeys. Marian Diamond and her associates (1981) demonstrated anatomical asymmetry in the cortex of the rat. Structural asymmetry is also known to be present in the lamprey, the frog, and other species.

The discovery of asymmetry in the brains of adult humans soon led to investigation of asymmetry in utero. LeMay found that asymmetry of the sylvian fissures did in fact develop during intrauterine life. J. G. Chi and associates (1977) showed that asymmetry of the temporal speech region also developed in utero. In addition, they made the important observation that the smaller homologous area on the right side develops significantly earlier than the larger left-sided region.

Although the very first studies on the development of anatomical asymmetry concentrated on fetal life, more recent research by Arnold Scheibel (see Chapter 4) on the maturation of the dendritic arborizations in the frontal language area of Broca strongly suggests that the final pattern of dominance can be modified in the first few years of postnatal life. Scheibel's data are also compatible with the view that the right-sided systems tend to mature earlier, only to be overtaken eventually by the left side. The studies of Patricia Goldman-Rakic (Chapter 12) have added yet another dimension to the study of intrauterine development of dominance. She has shown that lesions placed in one side of the cortex of the monkey in utero lead to the development of bilateral connections from the spared homologous region, and to enlargement of cortical regions in both the same and the opposite hemisphere, which she relates to loss of fewer neurons in the spared areas at the time of prenatal cell death.

Another major contribution was the discovery of chemical and pharmacological asymmetry in the brain. The possibility that certain drugs might act asymmetrically in the brain had been raised only rarely in the past. It was the work of Stanley Glick and collaborators (1977) that established the presence of this type of lateralization in the brain of the rat. Ida Gerendai has now demonstrated distinct asymmetries in the control of the ovary and other organs by structures in the central nervous system (see Chapter 11).

This continued progress has made it possible to investigate evolution of cerebral dominance, a problem which only 15 years ago appeared to be totally beyond the range of any available technique. LeMay, however, showed that certain asymmetries in the brain leave their impression on the skull. It has thus become possible to trace the evolution of brain asymmetries in humans and other primates back at least 300,000 years.

Radiological methods opened the way for consideration of other means of detecting brain dominance in the living human. LeMay, who had first investigated asymmetries seen on arteriography, initiated use of the completely noninvasive computerized radiological scanning methods for visualization of anatomical asymmetries. Other imaging methods make possible a dynamic picture of dominance. Thus methods for measuring blood flow and the more recent techniques for studying metabolism by injection of positron-emitting substances have made it possible to view asymmetrical changes in activity. The electroencephalogram has been important in the investigation of neurological patients for over 40 years, yet it has only recently been applied to dynamic studies of dominance. Advanced techniques such as Frank Duffy's BEAM method (see Chapter 5) permit study of the entire pattern of electrical changes across both hemispheres during the performance of different types of activity, as well as in different disease states in which dominance is disordered.

It had long been argued by investigators such as Samuel Torrey Orton, who was the first to attract widespread attention to the syndrome of childhood dyslexia, that this condition and many other learning disabilities were characterized by an anomalous pattern of cerebral dominance. Yet Orton himself, and many who followed him, believed firmly that this pattern had no structural correlate. The description by Galaburda and Thomas Kemper (1979) of neuronal migration defects in the speech region of the left hemisphere of a severe childhood dyslexic, and the finding of other cases with similar structural alterations, have opened a new field—the developmental neuropathology of cerebral dominance.

Even though dominance was implicitly recognized as a biological trait, it should be obvious from the above discussions that many beliefs which made cerebral dominance a most atypical biological trait persisted over very long periods. It was almost as if the common views concerning dominance reflected a dualistic view of the mind and the body. If dominance had no structural correlate and if it existed only in humans, then it was almost reasonable to consider it as the pure property of some hypothetical mental organ. As a result, the possibility that it had noncognitive biological associations was not taken seriously. Indeed, its position in medicine reflected the common view of dominance as an attribute of cognitive function only. Even within neurology it was viewed as a rather exotic subject, of little or no interest to the average neurological practitioner and of significance only for that small group with an interest in esoteric subjects such as aphasia.

Earlier studies pointing out some biological associations of dominance had aroused little broad interest. Certain investigators had called attention to the different patterns of fingerprints in right-handers and left-handers, others to the alteration in left-handers of the usual pattern of the testes (in which the right is higher and heavier). Several investigators had manifested interest in the hereditary patterns of handedness, but only recently have more fundamental biological considerations begun to penetrate studies of the genetics of laterality. Older studies revealed unexpected relationships between twinning and left-handedness. Charles Boklage has explored this subject extensively and has discussed its possible biological foundations (see Chapter 13). The medical significance of dominance has been demonstrated by Geschwind and Peter Behan (1982), who have documented a markedly elevated frequency of immune diseases, migraine, and other conditions in strongly left-handed people.

The techniques of investigation of dominance have continued to change over the years. The method first used by Broca, the clinical and postmortem study of patients with delimited unilateral brain lesions, continues to be fruitful. Hugo Liepmann's discovery of the syndromes of the corpus callosum early in this century added the method of studying the isolated hemispheres, which after a long period of neglect has again become important. Early in this century another still very valuable approach was developed, that of stimulating the exposed human cortex in awake patients during neurosurgical operations. Juhn Wada added another powerful technique, that of temporary paralysis of a hemisphere by injection of a barbiturate into the internal carotid artery of one side. Unilateral electroconvulsive therapy has been used, although less extensively, to produce similar effects.

A major advance resulted from the introduction of techniques (such as the dichotic listening and tachistoscopic methods) to test cerebral dominance in intact normals by restricting the stimulus to one hemisphere. These methods tap the specialized unilateral processing of sensory stimuli, just as studies of hand usage reflect unilateral specializations for motor function.

The biological studies reported in this volume have relied on another group of investigative methods, some of them classical but unique in their application to the study of cerebral lateralization. Anatomical, radiological, and electrophysiological techniques and chemical determinations have been added to the older methods. Behavioral studies in animals, such as those of Nottebohm (1977) in birds and Denenberg (1981) in rats, have led to some of the most startling results. Unexpectedly, endocrinology and immunology appear to play a role in the determination of dominance, while conversely the dominance pattern may alter both endocrine and immune status. One may confidently predict that the techniques of molecular biology will shortly be applied to study the basic determinants of laterality.

What we believe today about dominance is very different from what we all believed only a few years ago. The succeeding chapters should make it obvious that a whole new branch of biology is coming into being, which will have major implications for the understanding of cognitive function, for the elucidation of mechanisms of disease, and for almost all other branches of biology and medicine.

References

Chi, J. G., Dooling, E. C., and Gilles, F. H. 1977. Gyral development of the human brain. *Ann. Neurol.* 1:86–93.

Denenberg, V. H. 1981. Hemispheric laterality in animals and the effects of early experience. *Behav. Brain Sci.* 4:1–49.

Diamond, M. C., Dowling, G. A., and Johnson, R. E. 1981. Morphological cerebral cortical asymmetry in male and female rats. *Exp. Neurol.* 71:261–268.

Galaburda, A. M., and Kemper, T. L. 1979. Cytoarchitectonic abnormalities in developmental dyslexia: a case study. *Ann. Neurol.* 6:94–100.

Galaburda, A. M., Sanides, F., and Geschwind, N. 1978. Human brain: cytoarchitectonic left-right asymmetries in the temporal speech region. *Arch. Neurol.* 35:812–817.

Geschwind, N., and Behan, P. 1982. Left-handedness: association with immune disease, migraine, and developmental learning disorder. *Proc. Natl. Acad. Sci. USA* 79:5097–5100.

Geschwind, N., and Levitsky, W. 1968. Left-right asymmetry in temporal speech region. *Science* 161:186–187.

Glick, S. D., Jerussi, T. P., and Zimmerberg, B. 1977. Behavioral and neuropharmacological correlates of nigrostriatal asymmetry in rats. In S. R. Harnad, R. W. Doty, L. Goldstein, J. Jaynes, and G. Krauthamer, eds., *Lateralization in the Nervous System.* New York: Academic Press, pp. 213–249.

LeMay, M. 1976. Morphological cerebral asymmetries of modern man, fossil man, and nonhuman primate. *Ann. N.Y. Acad. Sci.* 280:349–366.

LeMay, M., and Culebras, A. 1972. Human brain: morphologic differences in the hemispheres demonstrable by carotid arteriography. *New Eng. J. Med.* 287:168-170.

LeMay, M., and Geschwind, N. 1975. Hemispheric differences in the brains of great apes. *Brain Behav. Evol.* 11:48-52.

LeMay, M., and Kido, D. K. 1978. Asymmetries of the cerebral hemispheres on computed tomograms. *J. Comp. Assist. Tom.* 2:471-476.

Nottebohm, F. 1977. Asymmetries in neural control of vocalization in the canary. In S. R. Harnad, R. W. Doty, L. Goldstein, J. Jaynes, and G. Krauthamer, eds., *Lateralization in the Nervous System.* New York: Academic Press, pp. 23-44.

Part One

Brain Asymmetry
in Humans

Chapter 2
Anatomical Asymmetries

Albert M. Galaburda

The primary goal of studies on brain asymmetry, even of those dealing with asymmetry of animal brains, is the better understanding of cerebral dominance in humans and of the mechanisms underlying abnormality of functional lateralization. During the years following the discovery by Paul Broca (1861) that lesions producing language disorders were generally located in the left hemisphere, many investigators endeavored to demonstrate structural differences between the two sides. More than a century after Broca our knowledge about the asymmetrical areas of the human brain is still incomplete, but asymmetry in several areas having to do with language representation has been demonstrated. Interestingly, anatomical asymmetries in animal brains were discovered years before functional lateralization was shown to be present, the converse of the situation in humans.

The significance of human cerebral asymmetries inspired the study of other asymmetries and their relation to disease processes. Chemical asymmetries are now known for some anatomically asymmetrical structures (Oke et al., 1978; Amaducci et al., 1981). Studies have been published on the relation of asymmetry to severity of aphasia (Pieniadz et al., 1979) and developmental dyslexia (Hier et al., 1978), although Hier's findings have been contested. We also have a clearer understanding of the relationship between asymmetry and handedness (Hochberg and LeMay, 1974) and lateralization for language (Ratcliff et al., 1980).

Gross Anatomical Asymmetries

Initial studies of structural asymmetries focused on parameters such as hemispheric weight and volume, proportions of gray and white matter, cortical thickness, and differences in cortical folding on the two sides. By and large, the asymmetries found were small or inconsistent, although repeating this research with modern methods may be productive. Later papers, largely ignored, did cite reproducible asymmetries. Eberstaller (1884) and Cunningham (1892) described asymmetries in the sylvian fissures, finding a consistent difference in the shape of the posterior ends of the two sides. The right fissure tends to curl up posteriorly, whereas the left proceeds more horizontally and reaches farther back (Fig. 2.1). Sylvian asymmetries have been confirmed by LeMay and Culebras (1972) and by Rubens et al. (1976). LeMay and her colleagues have also shown that the sylvian asymmetry is visible in cerebral arteriograms (LeMay and Culebras, 1972) and that the distribution of asymmetry varies with handedness (Hochberg and LeMay, 1974). A bias toward a higher right sylvian fissure is much stronger among strong right-handers than among left-handers, who in turn tend to have more symmetrical sylvian fissures.

The greater length of the left sylvian fissure leads to the supposition that the temporal and parietal opercula, which make up the floor and roof of the posterior portion of the sylvian fossa, are larger on the left. The presence of a larger left temporal operculum was first shown by Pfeifer (1936), who reported that a portion known as the planum temporale (a triangular region lying caudal to the transverse auditory gyrus of Heschl on the superior surface of the temporal lobe), was larger on the left side (Fig. 2.2). In a later study involving many autopsy specimens Geschwind and Levitsky (1968)

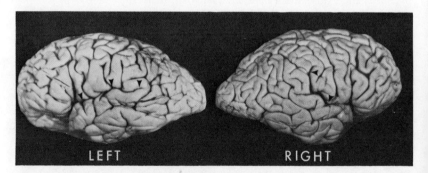

Fig. 2.1 The left and right cerebral hemispheres of the human brain, showing the most typical course of the sylvian fissures (arrows). Growth of the left parietal operculum may be accompanied by downward displacement of the sylvian fissure, while growth of the parietooccipital cortex on the right may result in anterior displacement of the caudal end of the sylvian fissure.

Fig. 2.2 The upper surfaces of the temporal lobes (supratemporal planes), illustrating the most typical type of asymmetry in the planum temporale (PT) — namely, larger on the left side. The arrows denote the posterior (inferior in figure) borders of PT. Note that the right PT, which lies posterior to H_2, is extremely narrow. A doubling or bifurcation of Heschl's gyrus (H) into two rostrocaudal segments (H_1 and H_2) is more common on the right side.

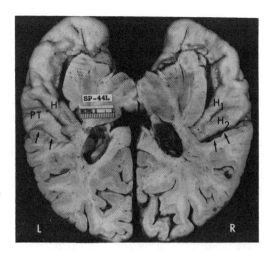

found that the planum temporale was larger on the left side in 65% of the brains, approximately equal in 24%, and larger on the right side in 11%. This distribution of planum asymmetry has since been confirmed by Wada (1969), Teszner et al. (1972), Witelson and Pallie (1973), Wada et al. (1975), Kopp et al. (1977). This portion of the temporal lobe contains high-order auditory association cortex (Galaburda and Sanides, 1980; Leinonen et al., 1980; Galaburda and Pandya, 1982) and is a typical site of injury in patients with Wernicke's aphasia (Meyer, 1950). These findings suggest that the asymmetry of the planum temporale is an important aspect of the anatomical lateralization of language to the left hemisphere. Yeni-Komshian and Benson (1976) have found a similar asymmetry in the chimpanzee.

Pfeifer also confirmed Heschl's statement that the right superior temporal plane was more likely to contain a second transverse gyrus (Fig. 2.2). This was subsequently reconfirmed by Beck (1955) and by Campain and Minckler (1976). Chi and coworkers (1977) showed additionally that the right planum temporale is more likely to be accompanied by two transverse gyri in utero.

The asymmetry of the planum temporale is often striking; it is not uncommon for the left planum temporale to be 10 times larger than the right. In some cases the right planum is virtually absent (Fig. 2.2). On the other hand, when the right temporal planum is larger, the asymmetry is usually much less marked. Although asymmetries of this magnitude can be seen in subcortical structures in other species such as the habenular nucleus of the lamprey (Braitenberg and Kemali, 1970), no other species has been shown to possess such a degree of cortical asymmetry — which may well be a unique feature of the human brain.

Broca was the first to give detailed accounts of the anatomy of the opercular portion of the frontal lobe. Many of the structures found there still carry his name. The cape of Broca (known also as the pars triangularis) is the middle section of the frontal operculum, bound anteriorly by the horizontal branch of the sylvian fissure and posteriorly by the ascending limb. Eberstaller (1884) commented that the ascending sylvian limb is branched more often on the left than on the right. The branching takes the shape of a diagonally oriented sulcus piercing the substance of the foot of Broca (pars opercularis) (Fig. 2.3). This suggested that the left posterior frontal opercular region contains more cortex buried in the folds, a suggestion recently supported by Falzi and colleagues (1982) who, in a study of 12 brains of right-handed individuals, found an average of 22% more infolded cortex in the left frontal opercular region. Mellus (1911) had pointed out that the left frontal operculum contains thicker cortex than the right. This may, in fact, reflect more extensive folding of cortex, which tends to increase cortical thickness especially at the peaks of gyri (Bok, 1959).

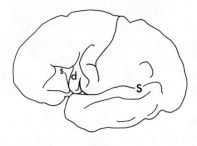

LEFT

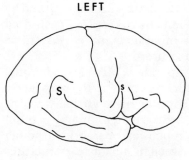

RIGHT

Fig. 2.3 Left and right cerebral hemispheres, showing a common pattern of asymmetry in the folding pattern of the frontal opercular region. The ascending limb (s) of the sylvian fissure (S) has a diagonal branch (d) more often on the left. This additional fold may account for the greater amount of opercular cortex on the left in this region, as shown by Falzi et al. (1982).

Additional gross anatomical asymmetries are present in the brain and can be seen in radiological investigations (see Chapter 3). Some have been related to asymmetries of the brain at postmortem (Pieniadz et al., 1981), some to the findings in East African skulls in which the right frontal and left occipital regions of the brain were seen to protrude beyond their mates (Gundara and Zivanovic, 1968). These are known as petalias, and the combination of right frontal petalia and left occipital petalia is the most common. The same brains often exhibit right midparietal petalia, although the sidedness of this petalia appears to be independent of the right frontal and left occipital combination, and may reflect asymmetries in functionally distinct regions of the brain.

The distribution of planum asymmetries—about 65% favoring the left and 35% roughly equal or favoring the right—appears at first glance inconsistent with the apparent lateralization of language dominance to the left hemisphere. It has been argued that the percentage of brains showing a larger left planum temporale is too low compared with the percentage of patients with aphasia after left-hemisphere lesions (Gainotti et al., 1982). However, it is quite likely that the 35% with equal plana or a larger right planum will typically have anomalous dominance for language and therefore absence of strong left-sided preponderance. These subjects, therefore, are more likely to be left-handed or ambidextrous and may become aphasic from lesions in either hemisphere. Luria (1970) found that almost all patients with penetrating wounds in the primary speech areas of the left were aphasic at onset. At a year, however, about 30% had made a good to excellent recovery and this group contained a very high proportion of left-handers and of right-handers with left-handed relatives. It is possible that the 35% without a larger left planum constituted the bulk of the group, which recovered better after left-sided lesions. Pieniadz and associates (1979) have reported that those individuals with atypical asymmetries on the CT scan showed somewhat better recovery from aphasia. It appears probable, then, that the number of subjects with anomalous language representation is larger than previously thought.

Architectonic Asymmetries

The search for asymmetries in the brain during the second half of the nineteenth century was restricted, mainly because of technical limitations to comparisons at the gross anatomical level. Demonstration of microscopic asymmetries had to await the development of better fixation, sectioning, and staining methodologies, which did not become available until the turn of the present century. With the emergence of these techniques, cerebral architectonics and quantitative microscopic study of regional variation were possible for the first time. The surface of the brain shows striking variation in the organization of the cortex. Cytoarchitectonics is the study of cellular organization. At low magnifications it is possible, in various regions of the cortex, to see differences in lamination, cellular size, cell packing densities, and distribution of cell types. The frontal cortex, for instance, tends to be thicker than the cortices of the temporal, parietal, and occipital lobes, and contains a higher proportion of large pyramidal neurons. In the frontal lobe an area known to give rise to fibers innervating the motor neurons of the spinal cord, the cortical motor area, exhibits many large pyramidal neurons and virtually no small granular cells. On the other hand, sensory cortices such as the primary visual, somesthetic, and auditory cortices are richly packed with small granular neurons and exhibit fewer large pyramids. There is strong evidence to support the notion that distinct architectonic areas have distinct connectional patterns and functions (Sanides, 1970). In sections stained for nerve fibers rather than cells, one also finds many regional variations; the study of these is known as myeloarchitectonics.

It is usually possible to specify the borders of architectonic areas and, in serially sectioned material, to measure the volume of architectonic zones. Of special interest in the study of asymmetry have been the architectonic areas thought to participate in language function. Early attempts at measurement produced conflicting data. The myeloarchitectonic study of Strassburger (1938), for instance, concluded that asymmetries in architectonic volumes of the frontal opercular regions were inconsistent. Economo and Horn (1930), on the other hand, showed that auditory association cortices were consistently larger on the left side. Most of these studies were done on preselected blocks, which made assumptions about the topography of areas and did not include the full extent of the relevant regions. By means of the application of the Vogt method (Sanides, 1970) we were able to parcellate the full extent of the auditory representations in the cortex of both hemispheres (Galaburda and Sanides, 1980). In a study aimed at specifying the relationship between the known planum asymmetry and the asymmetry of specific auditory association cortices, we found a consistent asymmetry in a region known as Tpt (Fig. 2.4). Area Tpt occupies a major portion of the planum temporale and adjacent posterior superior temporal gyrus. It is an area commonly affected in Wernicke's aphasia

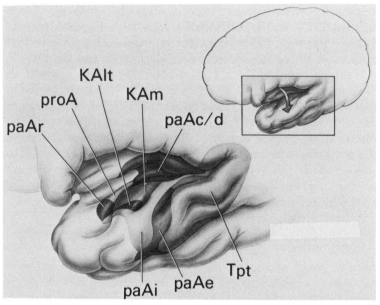

Fig. 2.4 Semistylized diagram of the superior temporal region of the human brain, which houses the auditory representations. Of the various architectonic areas Tpt shows the most consistent left-right asymmetries that parallel asymmetries in the planum temporale. It is located in that portion of the superior temporal region most often damaged in Wernicke's aphasia. The other labeled areas designate other cytoarchitectonic subdivisions that show little or no asymmetry.

(Meyer, 1950); connectional studies in human and monkey brains suggest that it is connected with inferior prefrontal regions in the frontal operculum (Galaburda and Pandya, 1982), an organization of connections thought to be important for language processing (Geschwind, 1970). Our findings showed that the left area Tpt is larger in brains that show a larger left planum temporale (Galaburda et al., 1978). Amaducci et al. (1981) have shown asymmetries in content of choline acetyltransferase in area Tpt, with a preponderance of this enzyme on the left side.

The posterior asymmetry in the sylvian fissure reflects in part the asymmetry of the planum temporale, but also an asymmetry in the inferior parietal lobule. A recent architectonic analysis of this region (Eidelberg and Galaburda, in press) demonstrated an asymmetry of area PG in favor of the left side. Area PG lies mainly on the angular gyrus. It is an evenly laminated (homotypical) cortex that matures late and is developed most fully in the human species.

Physiologically this cortex contains multimodal sensory neurons (Hyvärinen, 1979) and is connected to the frontal speech regions (Pandya and Vignolo, 1969) and to the language-relevant posterior thalamus (Van Buren and Borke, 1972; Eidelberg and Galaburda, 1982). Lesions involving the left angular gyrus often result in anomic aphasia and in reading and writing disturbances (Benson, 1979). Nucleus LP was found to be larger only in brains with a larger left planum temporale. This determination, together with the asymmetry in area Tpt in brains with a larger left planum, suggests that the asymmetry in language-related cortices may be linked. On the opposite side, an asymmetry present in the inferior parietal lobule in favor of the right side (that of area PEG) is not linked to the language-area asymmetries, but is perhaps related to the right midparietal petalia of Gundara and Zivanovic (1968).

The greater degree of folding of the left frontal operculum (Eberstaller, 1884; Falzi et al., 1982) may reflect an asymmetry in an architectonic field present in the pars opercularis of the inferior frontal gyrus, a region affected by lesions producing Broca's aphasia. Several architectonic fields are distinguished on the pars opercularis and pars triangularis of the frontal lobe (Braak, 1979). One of these has a striking appearance in sections stained for lipofuscin pigment (a technique known as pigmento-architectonics) because of the presence of large uniquely staining pyramids in layer IIIc. It takes up most of the volume of cortex present caudal to the diagonal sulcus of the ascending limb of the sylvian fissure; an architectonically homologous region in the monkey located in the caudal bank of the inferior limb of the arcuate sulcus receives the main body of projections from the posterior third of the superior temporal gyrus (Galaburda and Pandya, 1982). Furthermore, a physiological study by Rizzolatti et al. (1981a,b) has demonstrated the presence of multimodal sensory motor cells in this location. In an analysis of 10 brains serially sectioned in this region (Galaburda, 1980), the left opercular architectonic area was found to be more than 15% larger than the right in 6 brains, roughly equal in 3 brains, and larger on the right side in 1. In 3 of the brains with a larger left opercular area, the left side had more than twice the volume of the right. This architectonic asymmetry in the pars opercularis matches the asymmetry in folding described by Eberstaller (1884) and Falzi et al. (1982).

Lesion and stimulation data have shown that certain regions of the posterior thalamus (usually on the left side) are involved in language function (Bell, 1965; Ojemann, 1974). The pulvinar is the nucleus most often implicated, but adjacent mediodorsal and lateralis posterior nuclei are thought to participate also. The pulvinar is a large nucleus with multiple subdivisions difficult to delineate, and no consistent pulvinar asymmetries have been demonstrated. On the other hand, Eidelberg and Galaburda (1982) have shown that in 8 out of 9 brains the lateralis posterior nucleus is larger on the left. This nucleus is known to project to the inferior parietal lobule (Van Buren and Borke, 1972), which is consistent with the notion that it may also participate in language function.

The finding of variable degrees of architectonic asymmetry in language areas may support the claim that these anatomical differences underlie variations in the extent of functional lateralization for language. In Broca's area and in Wernicke's area, for instance, architectonic asymmetries vary markedly. In one brain the left area Tpt was 7 times larger than the right. In Broca's area architectonic asymmetries fluctuate between 15% and 259% in favor of the left side. It would not be surprising to discover that arguments about the effects of small lesions in the frontal operculum (Mohr, 1973) can be resolved by exact knowledge about the extent of anatomical asymmetry in that region. Even in the small study on architectonic asymmetries in Broca's area, two-thirds of the brains were found to have a larger left area, whereas the rest had either equal frontal opercula or a larger one on the right. These proportions are in keeping with the distribution of asymmetries found in the planum temporale, areas Tpt and PG, and in radiological studies (see Chapter 3); they again suggest the presence of anomalous dominance in roughly 35% of the population.

Asymmetries in the pathway to the parietal eye of the brain of the lizard (Engbretson et al., 1981) and in the architecture of the habenular nucleus (Braitenberg and Kemali, 1970) suggest that architectonic asymmetries may be accompanied by asymmetries in patterns of fiber connections. Such asymmetries in connections have been shown to be present in the human brain as well. Flechsig (1876) demonstrated in 40% of brains studied postmortem an asymmetry in the proportion of pyramidal fibers crossing from one side to the other in the decussations of the medulla. Yakovlev and Rakic (1966) showed asymmetry in the pyramidal decussations in newborn and fetal human brains. Kertesz and Geschwind (1971), in a study of 158 adult brains, found that 82% showed a more rostral decussation of the left pyramid to the right side. Yakovlev and Rakic also noted the more common occurrence of a relatively complete decussation from the left to the right and postulated that this finding is compatible with the left hemisphere's greater control over the right hand in most people than that of the right hemisphere over the left hand. Another asymmetry is detectable in the lower medulla (Smith,

1904); an aberrant circumolivary bundle deriving from the pyramidal tract is found on the left side more frequently than on the right. It is believed that the left aberrant bundle innervates the right facial nucleus, thus providing an anatomical substrate for control of some speech-related muscles on the dominant side.

It appears useful at this time to continue the study of asymmetries of fiber systems. Although the demonstration of asymmetries of cytoarchitectonic volume is a step in understanding the structural asymmetries underlying cerebral lateralization, the finding of asymmetry in connectional patterns would support the concept that lateralized function is handled by asymmetrical neural systems. It has been proposed that the region which is dominant for a given function has a larger number of connections to both sides of the nervous system than the corresponding nondominant area.

Chemical Asymmetry

Lateral asymmetries in the content of several transmitter substances have been demonstrated in the human brain. Oke et al. (1978) showed asymmetry in the content of norepinephrine in the thalamus. In the pulvinar the left hemisphere contains more norepinephrine, whereas the ventrobasal complex of the right side is richer in this neurotransmitter. Amaducci et al. (1981), as already mentioned, showed that Brodmann area 22 (part of which corresponds to area Tpt) contains greater choline acetyltransferase activity on the left side, and the degree of asymmetry increases in the posterior portion of the superior temporal gyrus, an area involved in language. Serafetinides (1965) administered LSD-25 preoperatively and postoperatively to 12 patients who underwent right temporal lobe removals and to 11 with left-sided ablation. He found that the typical perceptual responses to LSD disappeared after right, but not left, temporal lobectomy. One interpretation of these findings is that lateralized differences in response to a drug administered systemically is the result of an asymmetrical distribution of receptor sites or differences in sensitivity on the two sides. The occurrence of left-sided neurological signs in a chemically mediated illness such as depression (Brumback and Staton, 1981; Freeman et al., in press) is consonant with this concept.

Development of Cerebral Asymmetries

The asymmetrical postnatal development of the anterior language region in the human brain is the subject of Chapter 4. Some other features of asymmetrical development will be discussed here, with special reference to issues already raised. Asymmetry in the planum temporale has been noted in the brains of infants and fetuses (Witelson and Pallie, 1973; Wada et al., 1975; Chi et al., 1977). Some of the specimens examined by Galaburda (1978; Galaburda and Sanides, 1980) came from young individuals. In the parietal architectonic research of Eidelberg and Galaburda (in press) two of the brains came from children. The study showing preponderance of the thalamic left lateralis posterior nucleus (Eidelberg and Galaburda, 1982) includes brains of a 4-month-old male, a 4-year-old female, a 6½-year-old male, and an 11-year-old female. Brain asymmetries are present early in life and reflect biological characteristics that are at least in part independent of postnatal experience.

Asymmetry has also been demonstrated in the rates of development of the hemispheres. The emergence of cortical folding (of the gyri and sulci) apparently occurs earlier in the right hemisphere. Hervé (1888) pointed out that Broca's region develops earlier on the right in fetal life, and Fontes (1944) called attention to the earlier appearance of right perisylvian convolutions. Chi et al. (1977) have shown that the folding of the cortex of the right superior temporal plane proceeds faster than that of the left. In this study Heschl's gyrus was visible earlier on the right side, and in some cases right-sided folding occurred as much as two weeks ahead of left-sided. The left-hemisphere cortex surrounding the sylvian fissures illustrates the principle that slowly developing structures in the brain may ultimately become larger and better (Netley, 1980). Thus although the language regions of the left temporal lobe may develop more slowly, on average they ultimately reach a greater size and complexity of organization. The more prolonged growth period of these areas may, on the other hand, make them more vulnerable to disrupting influences. The brain findings in developmental dyslexia (Galaburda and Kemper, 1979) that show unilateral left-hemisphere cortical malformations originating during the period between 16 and 20 fetal weeks may be a consequence of this vulnerability.

Asymmetrical cortical malformations are also being demonstrated in the laboratory animal. The New Zealand Black mouse, which in adult life manifests a disease similar to lupus erythematosus, has been shown to develop malformations on one side of the cerebral cortex (Galaburda and Sherman, in preparation). The lesions of childhood dyslexia will be reviewed in detail in Chapter 6. The discovery of normal and abnormal brain asymmetry in nonhuman species will facilitate future study of the mechanisms leading to cerebral dominance and its breakdown in some pathological states.

Some of the work reported here was supported by NIH grant NS14018 and by grants from the Wm. Underwood Co., the Powder River Company, and the Essel Fund.

References

Amaducci, L., Sorbi, S., Albanese, A., and Gainotti, G. 1981. Choline-acetyltransferase (ChAT) activity differs in right and left human temporal lobes. *Neurology* 31:799-805.

Beck, E. 1955. Typologie des Gehirns am Beispiel des dorsalen menschlichen Schläfenlappens nebst weiteren Beitragen zur Frage der Links-Rechtshirnigkeit. *Dtsch. Z. Nervenheilk.* 173:267-308.

Bell, D. S. 1968. Speech functions of the thalamus inferred from the effects of thalamotomy. *Brain* 91:619-638.

Benson, D. F. (1979). Neurologic correlates of anomia. In H. Whitaker and H. Whitaker, eds., *Studies in Neurolinguistics.* New York: Academic Press, vol. 4, pp. 293-328.

Bok, S. T. 1959. *Histonomy of the Cerebral Cortex.* New York: Elsevier.

Braak, H. 1979. The pigment architecture of the human frontal lobe. I. Precentral, subcentral and frontal region. *Anat. Embryol.* 157:35-68.

Braitenberg, V., and Kemali, M. 1970. Exceptions to bilateral symmetry in the epithalamus of lower vertebrates. *J. Comp. Neurol.* 138:137-146.

Broca, P. 1861. Perte de la parole. Ramollisement chronique et destruction partielle du lobe antérieur gauche du cerveau. *Bull. Soc. Anthrop.* 2:219.

Brumback, R. A., and Staton, R. D. 1981. Depression-induced neurologic dysfunction. *New Eng. J. Med.* 305:642.

Campain, R., and Minckler, J. 1976. A note on the gross configurations of the human auditory cortex. *Brain and Lang.* 3:318-323.

Chi, J. G., Dooling, E. C., and Gilles, F. H. 1977. Gyral development of the human brain. *Ann. Neurol.* 1:86-93.

Cunningham, D. J. 1982. *Contribution to the Surface Anatomy of the Cerebral Hemispheres.* Dublin: Royal Irish Academy.

Eberstaller, O. 1884. Zur Oberflächenanatomie der Grosshirnhemisphären. *Wien Med. Blätter* 7:479, 642, 644.

Economo, C. v., and Horn, L. 1930. Uber Windungsrelief, Masse und Rindenarchitektonik der Supratemporalfläche, ihre individuellen und ihre Seitenunterschiede. *Z. Neurol. Psychiat.* 130:678-757.

Eidelberg, D., and Galaburda, A. M. 1982. Symmetry and asymmetry in the human posterior thalamus. I. Cytoarchitectonic analysis in normal persons. *Arch. Neurol.* 39:325-332.

Eidelberg, D., and Galaburda, A. M. Inferior parietal lobule: divergent architectonic asymmetries in the human brain. *Arch. Neurol.*: in press.

Engbretson, G. A., Reiner, A., and Brecha, N. 1981. Habenular asymmetry and the central connections of the parietal eye of the lizard. *J. Comp. Neurol.* 196:155-165.

Falzi, G., Perrone, P., and Vignolo, L. A. 1982. Right-left asymmetry in the anterior speech region. *Arch. Neurol.* 39:239-240.
Flechsig, P. 1876. *Die Leitungsbahnen in Gehirn und Ruckenmark des Menschen auf Grund entwicklungsgeschlichticher Untersuchungen.* Leipzig: W. Engelmann.
Fontes, V. 1944. *Morfologia do Cortex Cerebral.* Lisbon: Instituto de Antonio Aurelio da Costa Ferreira.
Freeman, R. L., Galaburda, A. M., Díaz-Cabal, R., and Geschwind, N. The neurology of depression: cognitive and behavioral deficits with focal findings in depression and resolution after electroconvulsive therapy. *Arch. Neurol.:* in press.
Gainotti, G., Sorbi, S., Miceli, M., and Amaducci, L. 1982. Neuroanatomical and neurochemical correlates of cerebral dominance: a minireview. *Res. Comm. Psychol. Psychiat. Behav.* 7:7-19.
Galaburda, A. M. 1980. La région de Broca: observations anatomiques faites un siècle après la mort de son découvreur. *Rev. Neurol. (Paris)* 136:609-616.
Galaburda, A. M., and Kemper, T. L. 1979. Cytoarchitectonic abnormalities in developmental dyslexia: a case study. *Ann. Neurol.* 6:94-100.
Galaburda, A. M., and Pandya, D. N. 1982. Role of architectonics and connections in the study of primate brain evolution. In E. Armstrong and D. Falk, eds., *Primate Brain Evolution: Methods and Concepts.* New York: Plenum Press, pp. 203-216.
Galaburda, A. M., and Sanides, F. 1980. Cytoarchitectonic organization of the human auditory cortex. *J. Comp. Neurol.* 190:597-610.
Galaburda, A. M., Sanides, F., and Geschwind, N. 1978. Human brain: cytoarchitectonic left-right asymmetries in the temporal speech region. *Arch. Neurol.* 35:812-817.
Geschwind, N. 1970. The organization of language and the brain. *Science* 170:940-944.
Geschwind, N., and Levitsky, W. 1968. Left-right asymmetry in temporal speech region. *Science* 161:186-187.
Gundara, N., and Zivanovic, S. 1968. Asymmetry in East African skulls. *Am. J. Phys. Anthrop.* 28:331-338.
Hervé, G. 1888. *La Circonvolution de Broca.* Paris: Delahage & Lecrosnier.
Hier, D. B., LeMay, M., Rosenberger, P. B., and Perlo, V. P. 1978. Developmental dyslexia. Evidence for a subgroup with a reversal of cerebral asymmetry. *Arch. Neurol.* 35:90-92.
Hochberg, F. H., and LeMay, M. 1974. Arteriographic correlates of handedness. *Neurology* 25:218-222.
Hyvärinen, J., and Shelepin, Y. 1979. Distribution of visual and somatic functions in the parietal association area of the monkey. *Brain Res.* 169:561-564.
Kertesz, A., and Geschwind, N. 1971. Patterns of pyramidal decussation and their relationship to handedness. *Arch. Neurol.* 24:326-332.
Kopp, N., Michel, F., Carrier, H., Biron, A., and Duvillard, P. 1977. Etude de certaines asymétries hémisphériques du cerveau humain. *J. Neurol. Sci.* 34:349-363.

Leinonen, L., Hyvärinen, J., Sovijarvi, A. R. A. 1980. Functional properties of neurons in the temporo-parietal association cortex of awake monkey. *Exp. Brain Res.* 39:203-215.

LeMay, M., and Culebras, A. 1972. Human brain: morphologic differences in the hemispheres demonstrable by carotid arteriography. *New Eng. J. Med.* 287:168-170.

LeMay, M., and Kido, D. K. 1978. Asymmetries of the cerebral hemispheres on computed tomograms. *J. Comp. Assist. Tom.* 2:471-476.

Luria, A. R. 1970. *Traumatic Aphasia.* The Hague: Mouton.

Mellus, E. L. 1911. A contribution to the study of the cerebral cortex in man. *Anat. Rec.* 5:473-482.

Meyer, A. 1950. *The Collected Papers of Adolf Meyer.* Baltimore: Johns Hopkins University Press.

Mohr, J. P. 1973. Rapid amelioration of motor aphasia. *Arch. Neurol.* 28:77-82.

Netley, C. 1980. Cognitive development, cerebral organization and the X chromosome. Presented at NATO, Advanced Studies Institute, Neuropsychology and Cognition, Augusta, Georgia.

Ojemann, G. A. 1974. Speech and short-term memory: alterations evoked from stimulation in pulvinar. In I. S. Cooper, M. Riklan, and P. Rakic, eds., *The Pulvinar-LP Complex.* Springfield, Illinois: Charles C Thomas, pp. 173-199.

Oke, A., Keller, R., Mefford, I., and Adams, R. N. 1978. Lateralization of norepinephrine in the human thalamus. *Science* 200:1411-13.

Pandya, D. N., and Vignolo, L. A. 1969. Inter-hemispheric projections of the parietal lobe in the rhesus monkey. *Brain Res.* 15:49-65.

Pfeifer, R. A. 1936. Pathologie der Hörstrahlung und der corticalen Hörsphäre. In O. Bumke and O. Foerster, eds., *Handbuch der Neurologie.* Berlin: Springer Verlag, vol. 6, pp. 523-626.

Pieniadz, J. M., and Naeser, M. A. 1981. Correlation between CT scan hemisphere asymmetries and morphological brain asymmetries of the same cases at post-mortem. Presented at the 19th annual meeting of the American Academy of Aphasia.

Pieniadz, J. M., Naeser, M. A., Koff, E., and Levin, H. C. 1979. CT scan cerebral asymmetry measurements and recovery in stroke patients with global aphasia. Presented at the 17th annual meeting of the American Academy of Aphasia.

Ratcliff, G., Dila, C., Taylor, L., and Milner, B. 1980. The morphological asymmetry of the hemispheres and cerebral dominance for speech: a possible relationship. *Brain and Lang.* 11:87-98.

Rizzolatti, G., Scandolara, C., Matelli, M., and Gantilucci, M. 1981a. Afferent properties of periarcuate neurons in macaque monkeys. I. Somatosensory responses. *Behav. Brain Res.* 2:125-146.

Rizzolatti, G., Scandolara, C., Matelli, M., and Gantilucci, M. 1981b. Afferent properties of periarcuate neurons in macaque monkeys. II. Visual responses. *Behav. Brain Res.* 2:146-163.

Rubens, A. B., Mahowald, M. W., and Hutton J. T. 1976. Asymmetry of the lateral (sylvian) fissures in man. *Neurology* 26:620-624.

Sanides, F. 1970. Functional architecture of motor and sensory cortices in primates in the light of a new concept of neocortex evolution. In C. R. Noback and W. Montagna, eds., *The Primate Brain: Advances in Primatology.* New York: Appleton-Century-Crofts, vol. 1, pp. 137-208.

Serafetinides, E. A. 1965. The significance of the temporal lobes and of hemispheric dominance in the production of LSD-25 symptomatology in man. *Neuropsychologia* 3:69-79.

Sherman, G. F., and Galaburda, A. M. 1982. Cortical volume asymmetry and behavior in the albino rat. *Neurosci. Abs.* 8:627.

Sherman, G. F., and Galaburda, A. M. 1984. Asymmetries in anatomy and pathology in the rodent brain. In S. D. Glick, ed., *Cerebral Lateralization in Subhuman Species.* New York: Academic Press, forthcoming.

Smith, G. E. 1904. A preliminary note on an aberrant circumolivary bundle springing from the left pyramidal tract. *Rev. Neurol. Psychiat.* 2:377-383.

Strassburger, E. H. 1938. Vergleichende myeloarchitektonische Studien an der erweiterten Brocaschen Region des Menschen. *J. Psychol. Neurol.* 48:477-511.

Teszner, D., Tzavaras, A., Gruner, J., and Hécaen, H. 1972. L'asymétrie droite-gauche du planum temporale — à propos de l'étude anatomique de 100 cerveaux. *Rev. Neurol.* 146:444-449.

Van Buren, J., and Borke, R. 1972. *Variations and Connections of the Human Thalamus.* New York: Springer-Verlag, vols. 1 and 2.

Wada, J. A. 1969. Interhemispheric sharing and shift of cerebral speech function. *Exc. Medica Intern. Cong. Ser.* 193:296-297.

Wada, J. A., Clarke, R., and Hamm, A. 1975. Cerebral hemispheric asymmetry in humans. *Arch. Neurol.* 32:239-246.

Witelson, S. F., and Pallie, W. 1973. Left hemisphere specialization for language in the newborn. Neuroanatomical evidence of asymmetry. *Brain* 96:641-646.

Yakovlev, P. I., and Rakic, P. 1966. Patterns of decussation of bulbar pyramids and distribution of pyramidal tracts on two sides of the spinal cord. *Trans. Amer. Neurol. Assoc.* 91:366-367.

Yeni-Komshian, G. H., and Benson, D. A. 1976. Anatomical study of cerebral asymmetry in the temporal lobe of humans, chimpanzees and rhesus monkeys. *Science* 192:387-389.

Chapter 3

Radiological, Developmental, and Fossil Asymmetries

Marjorie LeMay

Radiological studies have made it possible to demonstrate asymmetries in the brains of living humans and, further, have shown that such asymmetries are reflected in the shape of the skull. This finding in turn has led to the discovery of asymmetries in the skulls of fossil humans and of other primates. For the first time, then, it becomes possible to derive information concerning the evolution of cerebral dominance. Asymmetries in the brains and skulls of fetuses enable us to study the ontogeny of cerebral lateralization, and those also will be described in this chapter.

Radiological Findings in Living Humans

VENTRICULAR ASYMMETRIES Asymmetries in size of the ventricles of the brain have been observed radiologically for a number of years. Bruijn (1959), in an excellent monograph on the radiological findings of brain atrophy after pneumoencephalography (PEG), found the left lateral ventricle to be usually wider than the right. His data in 163 individuals show the midportion of the body of the lateral ventricles, the cella media (Fig. 3.1), to be wider on the left in 67%, wider on the right in 24%, and equal on the two sides in 9%.

Table 3.1 gives measurements of ventricular asymmetries in a personal series of PEGs of 80 males and 80 females over the age of 20 whose studies showed no evidence of localized ventricular enlargement or mass lesion. The anterior portion of the lateral ventricles, as measured by the septal-caudate distance on brow-up films, was found to be wider on the left in 72% and wider on the right in 14% ($\chi^2 = 62.67$, $df = 1$, $p < .001$). The tips of the temporal horns are smaller than the bodies of the ventricles and may vary slightly in

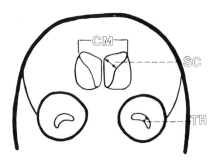

Fig. 3.1 Ventricles as seen on a pneumoencephalogram taken with the patient's forehead up (coronal view). CM = cella media, the width of the roof of the lateral ventricles; SC = septal caudate, distance between the superior margin of the septum pellucidum to the narrowest point of the margin of the caudate nucleus; TH = anterior portion of the temporal horn.

size on PEGs depending upon the amount of air in the horns, but in the same series of PEGs the tips of the temporal horns were found to be larger on the left in 68% and larger on the right in 11% ($\chi^2 = 62.87$, $df = 1$, $p < .001$). The differences in widths of the lateral ventricles in PEGs of young children were studied by Lodin (1968), who also found the left lateral ventricle to be usually wider than the right in children one year of age and older, but not in children younger than one year.

Asymmetries of ventricular size have also been observed on measurements of casts of the cerebral ventricles made after death. Such measurements of 200 fixed brains of individuals between the ages of 20 and 90 years showed the left lateral ventricle to be larger than the right in 48% and the right lateral ventricle to be larger in only 15%, with equality in 37% (Knudson, 1958). A similar study of casts of the ventricles of 21 fixed grossly normal human brains, done slightly earlier by Last and Thompsett (1953), also showed the left lateral ventricles to be usually larger than the right.

The variation in percentage of right-left differences between the PEG studies and the casts is not surprising. Ventricular size may be affected by shrinkage in brain size occurring with fixation of the brain postmortem, and also by agonal premortem events (Messert et al., 1972). Gas in the ventricles in PEG studies distends the ventricles, and ventricular differences can be seen more easily. Thus a higher proportion of equal measurements is to be expected when analyzing casts.

Ventricular measurement cannot be as accurately made at present on computerized tomography (CT) scans as on PEGs, but CT studies also show the tendency of the body of the left lateral ventricle to be slightly larger than that of the right. Gyldensted (1977), measuring the septal-caudate distance on CT scans of 100 normal adults, found the left to be wider in 62% of cases and the right wider in 32%, with a greater difference between the measurements on the

Table 3.1 *Comparative widths of the bodies of the lateral ventricles and tips of the temporal horns in 160 pneumoencephalograms (80 males and 80 females over the age of 20).*

Subjects	Septal caudate						Temporal horns					
	Right wider		Equal		Left wider		Right wider		Equal		Left wider	
	No.	%	No.	%	No.	%	No.	%	No.	%	No.	%
Males	11	14	10	12	59	74	13	16	14	18	53	66
Females	11	14	12	15	57	71	5	6	20	25	55	69
Total	22	14	22	14	116	72	18	11	34	21	108	68

two sides in males than in females. The brains and the ventricles are on average larger in males than in females. It is thus possible that some of the differences in widths were seen more easily on the CTs of males because more accurate measurements result when the ventricles are larger.

The size and shape of the lateral ventricles of the brain have been found to show some correlation with the size and shape of the brain and skull (Bailey, 1936; Berg and Lonnun, 1966). The most striking size differences occur in their posterior extensions, the occipital horns. Postmortem measurements of fetal brains show that the occipital horns occupy a larger proportion of the total ventricular volume than in later life; they are almost equal in size (Curran, 1909). In the seventh to eighth fetal month they almost reach the posterior ends of the hemisphere. The brain continues to grow posteriorly for a longer time than anteriorly. As the size of the brain increases even into adolescence, the occipital horns become relatively smaller and more asymmetrical in length, with the left occipital horn longer in most cases. This relative increase in length of the left occipital horn reflects the greater length of the left ventricle. Bruijn (1959) found the ventricles to be longer on the left in 77% of the PEG studies, longer on the right in 22%, and equal in 1%.

In 100 consecutive PEG studies of right-handers McRae et al. (1968) found the left occipital horn to be longer in 60%, the right longer in 31%, with equality in 9%. In another study in which the patients all had epilepsy, the same authors found the right and left occipital horns were of equal length in 60% of the right-handers and in 38% of the left-handers. These findings differ strikingly from the norm, suggesting that there was either abnormal brain development or trauma in fetal or postnatal life. Similar findings were reported in young children by Strauss and Fitz (1980): the left occipital horn was more often longer than the right, but in the PEGs of the children with seizures beginning before age 1 there was a much higher number with symmetrical or longer right occipital horns. Both studies show that abnormal development, or brain damage in early life which produces morphological changes, may shift the pattern of asymmetries.

VASCULAR ASYMMETRIES Studies of individuals undergoing diagnostic injection of radiopaque substances into the cerebral vessels (arteriograms) also commonly show differences of the position of the vessels on the two sides, which reflect morphological asymmetries of the hemispheres. The most striking lateral differences found by this method are in the posterior sylvian region.

In early fetal life there is an area, the insula, on the lateral surfaces of the hemispheres that grows more slowly than the adjacent brain. It is soon covered by portions of the frontal and parietal lobes from above, and the temporal lobe from below (the frontal, parietal, and temporal opercula). The opercula meet to form the sylvian fissure. The branches of the middle cerebral artery lie within the sylvian fissure and cross the surface of the insula. When the vessels are injected with contrast material, the area of the insula and the position of the sylvian fissures can be identified radiographically (Fig. 3.2).

In arteriograms of 106 right-handed patients without evidence of localized brain damage, the posterior end of the sylvian fissure (the sylvian point, SP) was higher on the right in 67%, higher on the left in only 8%, and equally high on both sides in 25% (Hochberg and LeMay, 1975). Szikla et al. (1975) reported similar findings. Comparable differences in height of the SP are seen in fetal brains (LeMay and Culebras, 1972; LeMay, 1976). In arteriograms of 28 left-handers the pattern was quite different: the SP was higher on the left in 21% and higher on the right in 8% (Hochberg and LeMay,

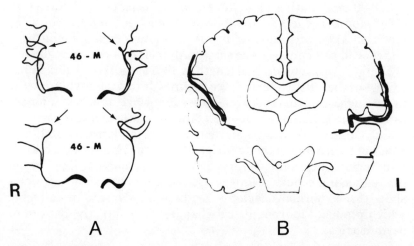

Fig. 3.2 A, *normal cerebral angiograms of two 40-year-old males, showing the course of the middle cerebral artery (MCA) on the left and right sides. At the posterior ends of the insula the angle formed by the branch of the MCA leaving the insular region (arrows) is wider on the right (R) than the left. B, coronal section through the brain at the posterior end of the sylvian fissures. The lower portion of the parietal lobe in this region is larger on the left than the right, and the MCA branches leaving the posterior end of the fissures are therefore lowered on the left (see Fig. 3.3).*

1975). The differences in height of the posterior ends of the sylvian fissures can be easily seen on the external surfaces of the brain (Fig. 3.3).

Ratcliff et al. (1980) reviewed arteriograms of patients in whom sodium amytal had been injected in the carotid arteries to determine which hemisphere was dominant for language (the Wada test). They excluded from the study patients who had a history suggesting possible brain damage before the age of 6 years. The SP was found to be higher on the right (as is the case with most right-handers) in individuals with left hemispheric speech, while patients with less typical speech representation showed less striking asymmetry.

Asymmetries sometimes are seen during the venous phase of arteriograms. One of the most predictable cerebral asymmetries is the turning to the right of the posterior end of the sagittal sinus. In the 20-mm fetus the major outflow of the superior venous channels was shown to be into the right transverse sinus (Streeter, 1975). Hochberg and LeMay (1975) found in 68 out of 111 (61%) carotid arteriograms that the major outflow of the sagittal sinus was into the right transverse sinus. In 19 (17%) the flow was to the left, and there was equal flow to the two sides in 22%. The transverse sinuses also usually lie at different heights. In a series of 97 arteriograms of right-handers (LeMay, personal observation) the left transverse sinus was found to be lower in 65 (69%) and the right to be lower in 26 (27%). In arteriograms of 29 left-handers the left was lower in 18 (62%) and the right lower in 5 (17%).

ASYMMETRIES SEEN BY COMPUTERIZED TOMOGRAPHY (CT) Cerebral asymmetries are readily seen by computerized X-ray studies, although reliable measurements are at times difficult to make because of technical problems such as angulation of the head in the scanner. Distortions may also appear in the photographic prints on which the measurements are made. Particularly when the image of the head appears unusually rounded on the photocopy, it is more difficult to compare the lengths and widths of the hemispheres.

CT studies show that the pineal calcification often lies slightly to the left of the midline and much less often to the right, demonstrating that the right hemisphere at the level of the pineal is commonly wider than the left (LeMay, 1976).

More striking asymmetries are seen at the ends of the hemispheres. CT studies (LeMay, 1976; LeMay and Kido, 1978) have shown that the anterior portion of the right hemisphere and the posterior portion of the left hemisphere are usually broader and extend farther than their counterparts. This has been found to be

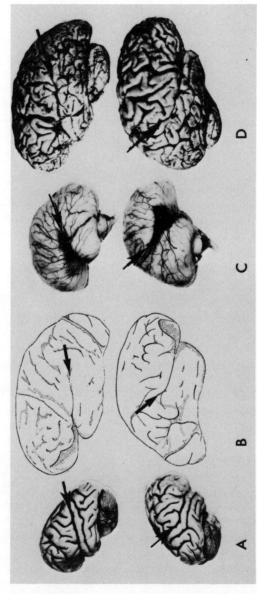

Fig. 3.3 Lateral views of the right (top row) and left cerebral hemispheres of A, an orangutan; B, an endocranial cast of Peking man (Risse et al., 1983); C, a human fetus; D, an adult male. Arrows mark the posterior ends of the sylvian fissures, which are higher on the right than on the left.

particularly true in right-handers. In left-handers there tends to be a higher percentage of individuals without striking asymmetries and with asymmetries which are the reverse of those most commonly seen in right-handers.

A few articles in the literature have reported an absence of significant differences between right-handers and left-handers in hemispheric asymmetries on CT studies. One of the studies was on children but 71% of them had or were suspected of having seizures (Deuel and Moran, 1980). As stated earlier, differences in the size of the occipital horns on PEG studies have been found in persons having epilepsy, particularly beginning in early life, when compared with those of individuals without seizures. Since a majority of the children in the above-reported CT study had had seizures, they may well have had brain abnormalities that influenced brain growth and hemispheric asymmetries on CT studies. In other research, headache patients have been used, but as the work of Geschwind and Behan (Chapter 14) shows, this group has an elevated rate of left-handedness. In some of the other studies that found no significant differences, the possibility of technical sources of error is being investigated.

To minimize personal bias in measuring hemispheric asymmetries, I have recently done a CT study on a Siemens scanner. Computerized measurements of hemispheric widths were taken from the console without knowledge of the handedness of the patients. The study group consisted of 185 right-handers and 44 left-handers without evidence of gross brain pathology causing any mass effect, who were competent mentally to be questioned about handedness. (I have found it is not reliable to take evidence of handedness from a patient's chart.) Table 3.2 shows the percentages of individuals whose cerebral hemispheres were wider on the right, on the left, or were of equal widths. The findings are similar to earlier studies in showing the anterior portions of the right hemisphere ($\chi^2 = 38.8$, $df = 1$, $p < .0005$) and the posterior portions of the left hemisphere ($\chi^2 = 60.0$, $df = 1$, $p < .0005$) to be wider than their counterparts in right-handers, with nonsignificant differences in left-handers.

Another reason for differences in measurements of brain asymmetries may be the populations being studied. Different ethnic groups have different facial appearances, and their head shapes also may vary. Risse et al. (1983) noted a pattern of cerebral asymmetries on CT studies in American Indians and blacks, different from that seen in whites. My own preliminary studies of CT scans of American blacks are in agreement with these findings; more brains show

Table 3.2 Comparative widths of frontal and occipital portions of the brain, as determined by CT scans.

Subjects		Frontal						Occipital					
		Left wider		Equal		Right wider		Left wider		Equal		Right wider	
Sex	No.	No.	%	No.	%	No.	%	No.	%	No.	%	No.	%
Right-handers													
Males	85	17	20	23	27	45	53	57	67	12	14	16	19
Females	100	18	18	17	17	65	65	72	72	13	13	15	15
Total	185	35	19	40	22	110	59	129	70	25	14	31	17
Left-handers													
Males	21	7	33	3	14	11	53	12	57	1	5	8	38
Females	23	7	30.5	9	39	7	30.5	7	30.5	7	30.5	9	39
Total	44	14	32	12	27	18	41	19	43	8	18	17	39

equal widths or a reversal of the hemispheric widths. Much larger numbers will be needed, however, to confirm or disconfirm these findings.

Hemispheric widths tend to correlate with the forward and posterior extension of the hemisphere (frontal and posterior petalia) on the same side, but the correlation is not perfect. I have found a slightly closer association between handedness and the widths of the hemispheres than between handedness and the petalias. Further data are needed to determine whether there is any relationship between handedness and cerebral asymmetries in a black population.

Asymmetries in lengths of the hemispheres as seen on skull measurement have been found in black and American Indian populations. Hrdlicka (1907) compared the relative lengths of the right and left sides of the cranial fossae in a series of adult, fetal, and children's skulls of white males and females, black males and females, and adult male American Indians (plus a few skulls of apes, monkeys, and other mammals). The skulls were from the College of Physicians and Surgeons in New York and the U.S. National Museum in Washington, D.C. The lengths of the frontal fossae were measured above the orbital plate of the frontal bone and those of the posterior fossae from the midpetrous ridge to the posterior margins of the inner table of the vault just above the groove of the lateral sinus. In all three ethnic groups the right frontal fossa and the left posterior fossa were longer than their counterparts much more often than the reverse was true. The asymmetries were in the same direction, but less striking, in the fetal and childhood skulls. Gundara and Zivanovic (1968) found asymmetries also in 98% of skulls of several groups of East African blacks at the Medical School at Makerere, Uganda. The most common asymmetries were forward extension of the right frontal bone and posterior extension of the left occipital region. Since most American blacks are of West African origin, it is conceivable that there are variations within different black groups.

CT scans of neonates tend to show a pattern similar to that of adults; that is, the anterior end of the right hemisphere and the posterior end of the left hemisphere are more often wider and protrude farther than their counterparts, at least in American whites (LeMay, 1977). The bones forming the vault are closely associated with the brain and dura (Young, 1959), and in fetal life they can be shown to move with the growing brain. One side, most often the right, of the foreheads of living individuals and of fetal skulls usually protrudes slightly farther forward. The coronal suture, and often the sphenoid wing, on the right side are also commonly for-

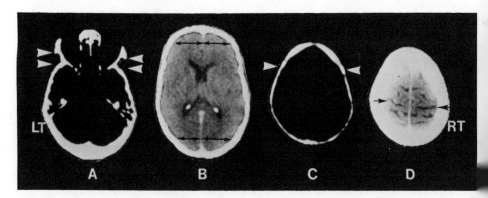

Fig. 3.4 Axial CT scans through the brain of a 70-year-old left-handed male. LT = left, RT = right. A, scan through the upper portions of the orbits and the middle posterior fossae. The upper arrows point to the outer rims of the orbits; the one on the left is more forward. The lower arrows show the level of the anterior margins of the sphenoid wings; the one on the left is more forward. B, scan through the ventricles. The anterior portion of the left hemisphere is wider and protrudes farther forward. The right hemisphere is wider posteriorly and on the scan section lying below this one protrudes farther posteriorly. The calcified glomus of the choroid plexus, and the thalamus just anterior to it, lie more posteriorly on the right. C, scan through upper portion of the vault. Arrows point to the coronal sutures. The left is more forward. D, scan at the same level as C. The outer margin of the central sulcus is more forward on the left (arrow).

ward of their counterparts (Fig. 3.4A; LeMay, personal observation). When the left forehead protrudes beyond the right, the coronal suture on the left is usually also forward (Fig. 3.4C), as well as the lambdoidal suture on the same side. I have seen this occasionally on CT, but the lambdoidal suture does not become as sclerotic as the coronal suture, since it closes later and is not seen as easily on CT scans.

The forward extensions on the same side of frontal lobe, forehead, and coronal suture are indications that the asymmetries of the brain are generally present in fetal life. The sulci, also formed in early life, can usually be easily identified on CT studies when there is sulcal widening. I have noticed that the central sulcus is often more anterior on the same side as the more forward coronal suture (Fig. 3.4D). Since the central sulci curve differently on the two sides, it is often difficult to judge accurately the anterior position of one relative to the other. There is also some variation in the position of the central sulci in the hemispheres; one would not expect the same close relationship between the central sulci and frontal petalia of the frontal lobes that one finds between the latter and the coronal suture. In excellent photographs by Fontes (1944), which have very little distortion, the central sulci of 29 fetal brains ranging from 4 to 9 gestational months — and also the parietal-occipital sulci — are much more frequently forward on the right than on the left. In 12 of the brains there was clear asymmetry in the position of the central sulci, and the one on the right was forward in 9 brains. The frontal petalias of these brains, which in some instances probably altered during fixation, were right in 3, left in 2, and equal in 4. In only 3 brains was the left central sulcus clearly forward, and all 3 showed a left frontal petalia.

Asymmetries of width and length of the hemispheres are most marked posteriorly. One can often see an indentation of the inner table of the occipital bone (possibly due to pressure from the brain during development), which is usually deeper on the left. There is commonly greater protrusion of the outer margin of the occipital bone on the left side than on the right (Smith, 1907). CT studies also show the thalamus and calcified glomus of the choroid plexus of the lateral ventricles to lie posterior to their counterparts in the same hemisphere that extends further posteriorly (LeMay, 1976; Fig. 3.4B). Hadziselimovic (1980), who has studied and written a great deal on postmortem cerebral and skull asymmetries, has shown a close relationship between the forward extension of the frontal horn of the lateral ventricle and frontal hemispheric petalia, and a longer occipital horn and posterior positioning of the splenium of the corpus callosum on the side of occipital hemispheric petalia. The cerebellum also often protrudes farther posteriorly on the same side of the hemisphere showing posterior petalia. This is perhaps caused by pressure during development from the cerebral hemisphere that is larger posteriorly, which leads to lowering of the tentorium and the cerebellum on the same side.

G. Elliott Smith, professor of anatomy at University College, London, in the early part of this century, argued for a relationship between head shape and handedness (Smith, 1907, 1908). Like other anatomists of the time, he believed that the reversal of the usual asymmetries (right frontal petalia and left occipital petalia) on the skulls he had examined occurred in almost the same proportion of skulls as left-handedness occurred in the population. Another argument was based on evidence that in right-handers the axial lengths of the right humerus and radius were greater than in left-handers. A colleague of Smith who measured skeletal bones in Nubia reported that in the majority of cases in which the cranial asymmetries were reversed, the bones of the left arm were longer (Smith, 1908). I have measured the lengths of both humeri in 77 right-handed and 17 left-handed women who were undergoing x-ray films for possible metastatic disease but found no statistical correlation between handedness and humeral lengths.

Fossil Evidence of Asymmetry

The presence of cerebral asymmetries in living humans leads us to realize that we can also study many of these asymmetries in ancient man and nonhuman anthropoids. The shape of the intracranial cavity is determined mainly by the shape of the brain; thus we can obtain data on the shape of the brain in ancient man by examining endocranial casts. There are difficulties even with this method, because most of the skulls found are fragmentary and many are filled with stone, which must be removed before endocasting can be attempted. In skulls of modern individuals, gyral impressions on the inner table of the vault are seen best in adolescence when the brain is still growing, or has only recently stopped. Connolly (1950) made many endocasts of skulls of modern humans and noted that although it was not always possible to see the entire extent of the sylvian fissures clearly, their posterior ends could usually be identified.

Drawings of the endocast of the skull of the Neanderthal man of La-Chapelle-aux-Saints, who lived approximately 50,000 to 60,000 years ago, show the posterior end of the sylvian fissure to be higher on the right than on the left, as is commonly seen in modern humans (Boule and Anthony, 1911; LeMay and Culebras, 1972). The same is true for the endocast of the skull of Peking man, who lived about 500,000 to 600,000 years ago (Shellshear and Smith, 1934; Fig. 3.3B). The shapes of the ends of the hemispheres are usually shown more clearly than the fissures and sulci of the brain. Holloway (1980, 1981), in studies of endocasts of skulls of early humans, has found forward extension of the right hemisphere and posterior extension of the left hemisphere beyond their counterparts to be the most common pattern of asymmetry.

Asymmetries similar to those in man are also seen in apes and, to a lesser degree, in some monkey brains (Holloway and De La Coste-Lareymondie, 1982; LeMay et al., 1982). Photographs of brains of great apes revealed asymmetries in the position of the sylvian fissures similar to those in humans (LeMay and Geschwind, 1981). In the brains of 28 higher apes, the posterior end of the right SP was higher in 16, the left was higher in only 1, with equally high ends in the remainder. These asymmetries were particularly prominent in the orangutan.

Endocranial casts of Old World and New World monkey skulls (most of them from the collection of D. Falk) and of some apes were studied (LeMay et al., 1982). Asymmetries in the ape endocasts resembled those in human skulls, with prominence of the right frontal and left occipital portions. Less striking asymmetries, but in a similar direction, were noted in the endocasts of Old World monkeys. Asymmetries were less common in the endocasts of the New World monkeys, but the posterior end of the left hemisphere protruded farther than its counterpart more often than did the right. Photographs and X-rays of the skulls showed the brow ridges of the apes to be usually more prominent on the right when the endocast of the same ape skull showed right frontal prominence (LeMay and Geschwind, 1981). One assumes that the basic shape of the skull, including the brow ridge, is mainly established in fetal life, as it is in humans (Fig. 3.5).

The most constant asymmetry of ape and monkey skulls and casts is in the major outflow of the superior sagittal sinus to the right and a lower position of the left transverse sinus (LeMay and Geschwind,

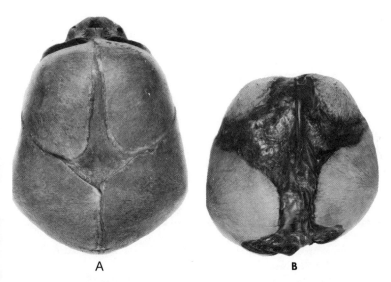

Fig. 3.5 A, fetal skull. The bone over the right frontal region and the coronal suture, forehead, and lower rim of the orbit are farther forward than on the left side. The vault extends slightly more posteriorly on the left. B, upper surface of skull of young fetus. The fetus probably had hydrocephalus, but again note the forward position of the right frontal region, the posterior extension of the left hemisphere beyond the right, and the positions of the bony islands of the developing vault.

1981). These changes appear to be at least as consistent in ape and monkey skulls as in those of humans.

The most striking finding, when one reviews cerebral asymmetries in ancient man and nonhuman anthropoids, is the increasing differentiation seen in the brains of modern individuals. This is not a new concept — it can be found nearly 200 years ago in Goethe's book *Metamorphosis*, which dealt with the science of comparative morphology (Riese, 1949). Goethe, presenting a pre-Darwinian rationalist theory of evolution, proposed that evolution leads to stricter definition of the parts and thus to a lack of symmetry and greater differentiation. Asymmetry was thus viewed as a sign of evolutionary progress that distinguished the simple from the complex and the primitive from the advanced. We are now learning, however, that asymmetries occur in lower forms. Their presence there and in fetal brains helps to substantiate the opinion of the early neuroscientist Pernkopf (see Keller, 1942) that morphological asymmetry finds expression in the early stages of ontogenesis.

References

Bailey, P. 1936. Variations in the shape of the lateral ventricles due to differences in the shape of the head. *Arch. Neurol. Psychiat.* 35:932.

Berg, K. L., and Lonnun, E. A. 1966. Ventricular size in relation to cranial size. *Acta Radiol. (New Ser. Diagnosis)* 4:65-78.

Boule, M., and Anthony, R. 1911. L'encéphale de l'homme fossile de la Chapelle-aux-Saints. *L'Anthrop.* 22:129-196.

Bruijn, G. W. 1959. *Pneumoencephalography in the Diagnosis of Cerebral Atrophy.* Utrecht, Netherlands: H. J. Smits Oudergracht.

Connolly, C. J. 1950. *External Morphology of the Primate Brain.* Springfield, Illinois: Charles C Thomas.

Curran, B. J. 1909. Variations in the posterior horn of the lateral ventricle, with notes on the development and suggestions as to their clinical significance. *Boston Med. Surg. J.* 161:777-782.

Deuel, R. K., and Moran, C. C. 1980. Cerebral dominance and cerebral asymmetries on computed tomograms in children. *Neurology* 30:934-938.

Fontes, V. 1944. *Morfologia do Cortex Cerebral.* Lisbon: Instituto de Antonio Aurelio da Costa Ferreira.

Gundara, N., and Zivanovic, S. 1968. Asymmetries in East African skulls. *Am. J. Phys. Anthrop.* 28:331-338.

Gyldensted, C. 1977. Measurements of the normal ventricular system and hemispheric sulci of 100 adults with computed tomography. *Neuroradiology* 14:183-192.

Hadziselimovic, H. 1980. Asymmetry of the human brain. *J. Hirnforsch.* 21:265-270.
Hochberg, F. H., and LeMay, M. 1975. Arteriographic correlates of handedness. *Neurology* 25:218-222.
Holloway, R. L. 1980. Indonesian "Solo" (Ngandong) endocranial reconstructions: some preliminary observations and comparisons with Neanderthal and Homo erectus groups. *Am. J. Phys. Anthrop.* 53:285-295.
Holloway, R. L. 1981. Volumetric and asymmetry determinations on recent hominid endocasts: Spy I and II, Djebel Ihrudi, and Solo Homo erectus. *Am. J. Phys. Anthrop.* 55:385-393.
Holloway, R. L., and De La Coste-Lareymondie, M. C. 1982. Brain endocast asymmetry in Pongids and Hominids: some preliminary findings on the paleontology of cerebral dominance. *Am. J. Phys. Anthrop.* 58:101-110.
Hrdlicka, A. 1907. Measurements of the cranial fossae. *Proc. U.S. Nat. Mus.* 32:177-201.
Keller, R. 1942. The asymmetry of the human body. *Ciba Symp.* 3:1126-27.
Knudson, P. A. 1958. *Ventriklernes Storrelsesforhold i Anatomisk normale Hjerner fra Voksne.* Copenhagen Theses, Odense, Denmark: Andelsbogtrykkeriet.
Last, R. J., and Thompsett, D. H. 1953. Casts of cerebral ventricles. *Brit. J. Surg.* 40:525-542.
LeMay, M. 1976. Morphological cerebral asymmetries of modern man, fossil man, and nonhuman primate. *Ann. N.Y. Acad. Sci.* 280:349-366.
LeMay, M. 1977. Asymmetries of the skull and handedness. *J. Neurol. Sci.* 32:243-253.
LeMay, M., and Culebras, A. 1972. Human brain: morphologic differences in the hemispheres demonstrable by carotid arteriography. *New Eng. J. Med.* 287:168-170.
LeMay, M., and Geschwind, N. 1981. Morphological cerebral asymmetries in primates. In B. Preilowski and C. Engele, eds., *Is There Cerebral Hemispheric Asymmetry in Non-Human Primates?* Tübingen: University of Tübingen.
LeMay, M., and Kido, D. K. 1978. Asymmetries of the cerebral hemispheres on computed tomograms. *J. Comp. Assist. Tom.* 2:471-476.
LeMay, M., Billig, M. S., and Geschwind, N. 1982. Asymmetries of the brains and skulls of non-human primates. In E. Armstrong and A. Falk, eds., *Primate Brain Evolution: Methods and Concepts.* New York: Plenum Press.
Lodin, H. 1968. Size and development of cerebral ventricular system in childhood. *Acta Radiol.* 71:385-392.
McRae, D. L., Branch, C. I., and Milner, B. 1968. The occipital horns and cerebral dominance. *Neurology* 18:95-98.

Messert, B., Wannamaker, B. B., and Dudley, A. W. 1972. Reevaluation of the size of lateral ventricles of the brain. *Neurology* 22:941–951.

Ratcliff, G., Dila, L. C., Taylor, L., and Milner, B. 1980. The morphological asymmetry of the hemispheres and cerebral dominance for speech: a possible relationship. *Brain Lang.* 11:87–98.

Riese, W. 1949. Goethe's conception of evolution and its survival in medical thought. A tribute on the occasion of the bicentenary of Goethe's birth. *Bull. Hist. Med.* 23:546–553.

Risse, G. L., Rubens, A. B., and McShane, D. 1983. Cerebral asymmetries on CT scan in three ethnic groups. Presented at the International Neuropsychological Society, Mexico City.

Shellshear, J. L., and Elliott Smith, G. 1934. A comparative study of the endocasts of Sinanthropus. *Phil. Trans. Royal Soc. London, Ser. B* 223:469–487.

Smith, G. 1907. On the asymmetry of the caudal poles of the cerebral hemispheres and its influence on the occipital bone. *Anat. Ann.* 30:574–578.

Smith, G. 1908. Right-handedness. *Brit. Med. J.* 2:596–598.

Strauss, E., and Fitz, C. 1980. Occipital horn asymmetry in children. *Ann. Neurol.* 8:437–439.

Streeter, G. I. 1915. The development of the venous sinuses of the dura mater in the human embryo. *Am. J. Anat.* 18:148–178.

Szikla, G., Hori, T., and Bouvier, G. 1975. The third dimension in cerebral angiography. In Y. Salamon, ed., *Advances in Cerebral Angiography*. Berlin: Springer-Verlag.

Young, R. W. 1959. The influence of cranial contents on post-natal growth of the skull in the rat. *Am. J. Anat.* 105:383–415.

Chapter 4

A Dendritic Correlate of Human Speech

Arnold B. Scheibel

The possibility of links between specific human talents and definable neural substrates was first brought to my attention in Neustadt in 1954 by Oscar and Cecile Vogt. From their large library of human brain tissue they showed me Nissl-stained brain sections from two individuals with documented talents: an artist with a lifetime capacity for eidetic imagery and a musician with absolute pitch. In the first individual, lamina IV of primary cortex (area 17) appeared twice as wide as that of ungifted controls. In the second, cortical lamina IV in Heschl's gyrus was at least twice the expected size. It seemed as if the number of cells in these cortical zones in the gifted individuals was not appreciably increased. The packing density of the cells, in fact, appeared lower, thereby providing the impression that enchanced laminar thickness might result, at least in part, from the presence of a more extensive neuropil. Because of my long interest in axonal and dendritic neuropil as visualized by Golgi methods (Ramon y Cajal, 1908-1911; Scheibel and Scheibel, 1954; Scheibel and Scheibel, 1955), I suspected that the increased laminar depth in these gifted individuals might reflect extended dendritic systems; increase in *size* of neuron somata, and in *number* of neuroglial cells, were distinct additional possibilities.

The possibility of performing quantitative studies on brain tissue of this type, using Golgi techniques, continued to be an intriguing one, despite the obvious difficulty of obtaining tissue specimens from those with unique intellectual powers. The studies of Geschwind (1968) in the mid-sixties and thereafter (Geschwind, 1974; Galaburda, Sanides and Geschwind, 1978) directed attention to possible correlates between functionally asymmetric faculties such as speech and morphological asymmetries between the hemispheres. The structural changes described were most evident on the superior surface of the temporal lobe (the planum temporale) and appeared to be concordant with the predominance of the language capacities of the left posterior cortical speech zone of Wernicke.

The thought occurred to me that since language function is located predominantly in one hemisphere, each individual, by virtue of possession of a very-much-less-speech-gifted hemisphere might serve as his/her own control. The existence of such an internal control would make it possible to study the substrate of a special talent in many brains.

Our initial attempts to compare the dendritic structure of neurons similarly situated in the two hemispheres were hampered by the extreme variability in gross cortical structure of the two hemispheres. A correlative electrophysiological-neuropsychological study of this problem in human neurosurgical subjects undertaken by one of our group (I. Fried) with G. Ojemann at Seattle (1981), confirmed the structural-functional variability of the language regions, but also indicated that the anterior speech area of Broca centered on the posterior part of the third frontal convolution might provide more reasonable opportunities than the posterior area of Wernicke for correlation between structure and function.

This chapter details our histological comparison of dendrite patterns of layer III pyramidal neurons in the left opercular (Broca's area) and right opercular regions (the less-speech-gifted "control" area on the nondominant side), and those portions of the adjacent left and right precentral gyri which control orofacial movements. Examination of these four areas, we hoped, would provide the opportunity for sampling cell systems involved in both the "strategies" (opercular) and the "tactics" (precentral) of speech function, as well as for exploring the impact of functional asymmetry on paired homologous cortical areas.

Our subject pool was composed of 8 recently deceased (within 6 to 18 hours) individuals whose ages ranged from 47 to 72 years and who had died of nonneurological causes. The fact that all our patients were males introduces a potential artifact of selection, particularly in view of studies such as those of Diamond (Chapter 9), which indicate significant male-female differences in degree of anatomical lateralization. Small tissue blocks (1 cm^3) were removed from the foot of left and right precentral gyri and from the opercular regions just anterior to them, to include portions of the small contiguous gyri known as the opercularis triangularis. Since this tissue block selection depended, in the final analysis, on proper identifi-

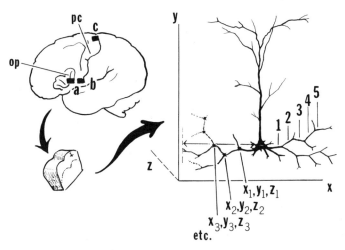

Fig. 4.1 Summary of the preparation of tissue for analysis. Starting from upper left, three tissue blocks (a, b, and c) are removed from each hemisphere—one from the opercular zone (op) and two from the precentral gyrus (pc). Each tissue block is fixed, impregnated, and sectioned, and the basilar dendrite system of each neuron selected is analyzed. Dendrite segment orders are numbered centrifugally 1, 2, 3, . . . Each bifurcation point ($X_1Y_1Z_1$, $X_2Y_2Z_2$, $X_3Y_3Z_3$, . . .) is identified by X and Y values stipulated by the camera lucida drawing and Z value determined by its precise plane of focus within the tissue.

cation of the precentral gyrus, we also took confirmatory blocks from the vertex region of each presumed precentral gyrus (Fig. 4.1). The giant pyramidal cells of Betz are found in the highest concentration in this area (Lassek, 1954) and serve as a positive anatomical marker for the motor strip.

Most of each block was stained by a variant of the rapid Golgi (osmic-dichromic-silver) method, while the remainder was stained for control use with cresyl violet. All slides were coded by the technician to assure observer impartiality.

As a working hypothesis we assumed that the more sophisticated levels of processing demanded by the development of linguistic strategies in the speech-dominant cortex (as compared to the non-speech-gifted cortex of the opposite side, or to the strictly "motor" areas just behind) would necessitate a more elaborate dendritic tree. For a number of operational reasons we decided to confine ourselves to the basilar dendrite systems of layer III pyramidal cells (defined as all cells of appropriate configuration lying at depths of 500 to 1,200 microns beneath the cortical surface). Such dendrite ensembles have a complex branching system made up of easily identifiable branching orders, 1, 2, 3, . . . , n. Six cells were selected in each of the four relevant areas — the left opercular (Lop), right opercular (Rop), left precentral (Lpc), and right precentral (Rpc) — thereby providing 24 cell samples per brain and 192 for the entire study.

Two separate methods were used for quantitative description of the cells. In the first phase of the study, a series of concentric rings etched in the eyepiece was superimposed on the cell-dendrite system, and the dendritic intersections with each ring were counted. This type of analysis, pioneered by Scholl in his studies of visual cortex (1956), provided a measure of complexity and branching pattern of the dendrite tree and, in retrospect, proved surprisingly useful for pilot studies when compared with our second and more elaborate computer-assisted analysis.

In the second phase of the study, each cell was drawn with the aid of a camera lucida at a magnification of 312.5X. The precise point of "best focus" was determined for the cell body; this value in microns, read from the calibrated fine focus of the microscope, became the reference depth. At each dendrite bifurcation the precise point of best focus was again determined and noted, and so on to the tip, thereby providing a measure of the extension through depth of each branch and a truer idea of dendritic length. (Each Golgi-stained section was 120 microns in thickness.) We were thus provided with "Z-axis" values to complement the X and Y values offered by the drawings themselves. Each drawing was then entered into a bit pad tied to a microcomputer, and all of the segment length and segment number values were punched into cards (approximately 10,000 of them) for evaluation by the medical center computer facility. No attempt was made to follow dendrite branches into adjacent sections, but tips were identified as "terminal," "broken," or "indeterminate."

Analysis of the data revealed significant differences in parameters of dendritic structure, several of which were unexpected. All of the data were originally treated in the same manner, without reference to the handedness of the individuals. Toward the end of the study we were finally able to track down the families of all of the patients (no handedness data were found on any of the charts) and thus discovered that six had been life-long right-handers, and two had been nonright-handed. After this information became available, we divided the data into two groups, those of the right-handers and those of the nonright-handers, since it is known that language dominance is either less clearly stated or is in some cases reversed for this second smaller group.

We were particularly interested in three measures: (a) the total dendritic length (tdl) of the basilar tree, as a measure of the total dendritic area available for processing operations; (b) the average segment length (asl), as an indication of the length achieved by a segment of a certain order; and (c) the average number of segments (asn) of a certain order, as a measure of the degree of "branchiness" achieved by the total dendritic arbor at that segment level.

Several findings proved of particular interest:

(1) Looked at as a whole, the tdl of opercular cells exceeded that of precentral cells by at least 10%, a difference that turned out to be significant ($p < .05$) only in the left hemisphere. This intuitively appealing difference might be attributed to the need for a more extensive axodendritic input system in cells in speech strategy (opercular) areas. (Although the nondominant side appears to be less capable of sustaining speech function beyond the first 8 to 12 years of life, it clearly continues to contribute significantly to overall patterns of linguistic behavior.)

(2) We were surprised to find that there were no significant differences between the tdl of basilar dendrite arbors of the right and left sides (left opercular, 2,714.5 mm vs. right opercular, 2,540.8 mm, nonsignificant; left precentral, 2,208.3 mm vs. right precentral, 2,499.9 mm, nonsignificant). This was particularly unexpected in the case of left and right opercular regions where we anticipated that the language-dominant side, carrying the great majority of speech strategy operations, would have shown a more extensive dendrite system.

(3) When we analyzed the dendrite arbors, segment order by segment order (Fig. 4.2), an interesting dichotomy emerged. On the right (nondominant) side, a larger fraction of the tdl was accounted for by lower-order (1,2,3) dendrite segments as expressed by the greater average lengths of the segments on this side (Table 4.1; Figs. 4.3 and 4.4). On the left (dominant) side, a larger fraction was accounted for by higher-order (4,5,6) dendrite segments as expressed by a larger number of segments of each of these orders. Interestingly, these values were partially reversed in the two nonright-handed patients. In one, the mean length of secondary and tertiary dendrites was greater in the left (compared to right) precentral cortex, whereas in the other, the mean length of tertiary dendrites was greater on the right but that of secondary dendrites was greater on the left. In the latter individual as well as in one of the right-handed cases, fourth-order dendrites were more numerous in the right operculum than in the left, again reversing the pattern found in most of our right-handed subjects.

Table 4.1 *Average segment length and mean number of dendritic branches by dendritic branch order, adjusted for depth by analysis of covariance.*

Cortex area	Dendritic branch order					
	1	2	3	4	5	6
Average segment length — mean (μm) ± standard error						
Lop	23.71 ± 3.17	49.74 ± 4.82	71.53 ± 5.03	86.4 ± 7.25	79.38 ± 7.90	88.86 ± 14.48
Rop	28.41 ± 3.16	55.57 ± 4.80	74.82 ± 5.01	87.50 ± 7.35	82.82 ± 8.21	93.23 ± 29.11
Lpc	28.51 ± 3.28	47.97 ± 5.03	62.13 ± 5.20	68.43 ± 7.60	86.87 ± 8.68	107.15 ± 24.53
Rpc	29.78 ± 3.18	70.15 ± 4.84	82.52 ± 5.04	82.24 ± 7.52	83.90 ± 8.50	64.99 ± 12.68
Main effects		$R > L$	$R > L$			
		$F = 8.21$	$F = 4.15$			
		$p < .005$	$p < .05$			
Average dendritic number — mean ± standard error						
Lop	5.20 ± 0.26	10.59 ± 0.56	12.57 ± 0.99	9.52 ± 1.38	3.72 ± .80	0.74 ± 0.18
Rop	5.17 ± 0.26	10.80 ± 0.56	12.55 ± 0.99	6.89 ± 1.37	2.58 ± .80	0.42 ± 0.18
Lpc	4.95 ± 0.27	10.10 ± 0.58	12.18 ± 1.01	7.72 ± 1.39	2.29 ± .82	0.09 ± 0.18
Rpc	5.24 ± 0.26	10.56 ± 0.56	11.33 ± 1.00	6.36 ± 1.37	1.49 ± .81	0.23 ± 0.18
Main effects				$L > R$		$op > pc$
				$F = 7.31$		$F = 4.24$
				$p < .01$		$p < .05$

A Dendritic Correlate of Human Speech 49

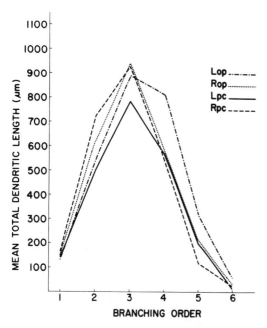

Fig. 4.2 *Mean total dendritic length for each segment order in cells of left opercular (Lop), right opercular (Rop), left precentral (Lpc), and right precentral (Rpc) areas of cortex. Lop and Lpc show shorter segment lengths through the third segment order, and starting with the fourth order branches Lop develops branches of greater length compared to the other cortical areas.*

In evaluating the significance of these results, we have tried to consider them in both a physiological and an ontogenetic context. Within the former frame, variations in dendritic structure and associated afferent terminal systems are known to occur depending on such factors as dendrite diameter, distance from the cell body, and bifurcation patterns. Thus, first-order dendrites are usually largest in diameter, have the smallest internal resistance values, and contain the fewest dendritic spines (Rall, 1962; Marin Padilla, 1967; Valverde, 1967). Correlated with this is a proportionately larger ensemble of Type II synaptic terminals, which may be predominantly inhibitory in function (Uchizono, 1965). Dendrite segments of progressively higher order are characterized by successively smaller diameters, higher internal resistance values, and greater dendritic spine density values (although these decrease again as the outermost segment orders are reached — Rall, 1962; Marin Padilla, 1967; Valverde, 1967). With the increased numbers of spines there are also increased numbers of Type I (probably facilitatory) synapses (Colonnier, 1968). Thus the left opercular cell,

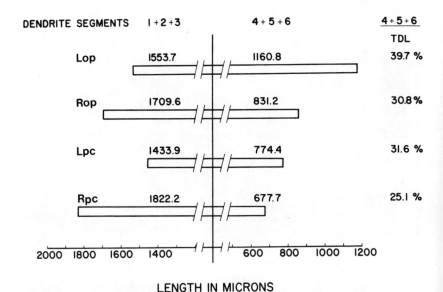

Fig. 4.3 Dendritic length and proportion of the dendritic ensemble made up of lower-order (1, 2, 3) and higher-order (4, 5, 6) dendritic segments in left opercular (Lop), right opercular (Rop), left precentral (Lpc), and right precentral (Rpc) areas. Comparing left against right, the approximate standard error is 138; comparing opercular against precentral areas, the approximate standard error is 107. The column of figures on the extreme right shows the percentage of tdl occupied by higher-order dendrites in each region.

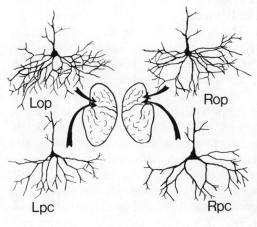

Fig. 4.4 Somewhat schematized drawing of typical dendritic ensembles from cells of the four described in the text. Arrows indicate the locations of the area studied. Note the increased number of higher-order segments in the left operculum (Lop) compared to all other areas, and the relatively greater length of second- and third-order branches in the right operculum (Rop) and right precentral (Rpc) areas.

with its greater number of higher-order dendrites and its shorter lower-order segments, may have functional characteristics that differ significantly from cells in the three other regions we have studied. In addition, since each dendritic bifurcation represents a "go-no-go" decision point for each region of local activity in the dendritic tree, more branchiness may be associated in a general way with greater degrees of freedom (and complexity) in information processing.

From the point of view of ontogeny, it must be remembered that there is a time-linked progression in the development of dendrite systems (Ramon y Cajal, 1908-1911). In the human neonate, only lower dendrite segments are present (usually orders 1 and 2). Systems in primary sensory and motor areas appear somewhat more advanced than those in secondary or association zones. The sequential Golgi studies of Conel (1943-1959) indicate that higher-order dendrite segments begin to appear after the sixth and twelfth month of postnatal life and become increasingly rich after the second year. If the dendritic patterns we have studied in older adults can be viewed as a fossilized history of earlier dendrite growth, it would seem that the longer lower-order segments on the nondominant side point to greater activity here in the prenatal and early postnatal (nonverbal) period of life, that is, the sensory-motor period of Piaget (1954). The marked increase in numbers of higher-order branches on the left side, relative to the right, coincides in time with the beginnings of conceptualization and speech function and may mark a nodal point in shift of dominance patterns from right to left.

At best, these data provide interesting correlations, but they are noncommittal about causality. We cannot yet specify whether the development of more complex higher-order dendrite patterns on the left provide a more suitable matrix for speech development or, alternatively, whether the onset of speech function already programmed preferentially for the left side stimulates increased dendritic branching. We have embarked on a companion study, trying to characterize quantitatively the development of dendrite systems in the precentral and opercular areas, left and right, from the perinatal period till the age of 4 to 5 years. This investigation may throw further light on the problem at hand.

References

Colonnier, M. 1968. Synaptic patterns on different cell types in the different laminae of the cat visual cortex. An electron microscope study. *Brain Res.* 9:268-286.

Conel, J. 1943-1959. *The Postnatal Development of the Human Cortex.* Cambridge, Mass.: Harvard University Press, vols. 1-6.

Fried, I., Ojemann, G., and Fetz, E. 1981. Language-related potentials specific to human language cortex. *Science* 212:353-356.

Galaburda, M., Sanides, F., and Geschwind, N. 1978. Human brain. Cytoarchitectonic left-right asymmetries in the temporal speech area. *Arch. Neurol.* 35:812-817.

Geschwind, N. 1974. The anatomical basis of hemispheric differentiation. In S. J. Dimond and J. G. Beaumont, eds., *Hemisphere Function in the Human Brain.* New York: Wiley.

Geschwind, N., and Levitsky, W. 1968. Left-right asymmetries in temporal speech area. *Science* 161:186-187.

Lassek, A. M. 1954. *The Pyramidal Tract.* Springfield, Illinois: Charles C Thomas.

Marin Padilla, M. 1967. Number and distribution of the apical dendritic spines of the layer V pyramidal cells in man. *J. Comp. Neurol.* 131:475-489.

Piaget, J. 1954. *The Construction of Reality in the Child.* New York: Basic Books.

Rall, W. 1962. Electrophysiology of a dendritic neuron model. *Biophys. J.* 2:145-167.

Ramon y Cajal, S. 1908-1911. *Histologie du système nerveux de l'homme et des vertébrés.* 2 vols. Paris: A. Maloine.

Scheibel, M. E., and Scheibel, A. B. 1954. Observations on the intracortical relations of the climbing fibers of the cerebellum. A Golgi study. *J. Comp. Neurol.* 101:733-764.

Scheibel, M. E., and Scheibel, A. B. 1955. The inferior olive. A Golgi study. *J. Comp. Neurol.* 102:77-132.

Scholl, D. A. 1956. *The Organization of the Cerebral Cortex.* London: Methuen.

Uchizono, K. 1965. Characteristics of excitatory and inhibitory synapses in the central nervous system of the cat. *Nature (Lond.)* 207:642.

Valverde, F. 1967. Apical dendritic spines of the visual cortex and light deprivation in the mouse. *Exp. Brain Res.* 3:337-352.

Witelson, S. F., and Pallie, W. 1973. Left hemisphere specialization for language in the newborn. Neuroanatomical evidence of asymmetry. *Brain* 96:641-646.

Chapter 5

Brain Electrical Activity Mapping

Frank H. Duffy
Gloria B. McAnulty
Steven C. Schachter

Functional specialization of the human cerebral hemispheres is widely accepted by neuroscientists and clinicians. Studies of penetrating head injuries, strokes, and tumors have clearly demonstrated the correlation between functional disturbance of language and a left-hemisphere locus of the lesion. Lesion definition has traditionally depended on the classic technique of gross and microscopic neuropathological examination of postmortem tissue. The obvious disadvantage is that this approach is not applicable to nonlethal conditions. There are, for example, only two published autopsies of patients with dyslexia (Drake, 1968; Galaburda and Kemper, 1979). The careful clinical-neuropathological correlations that are the underpinnings of adult neurology and neuropsychology are seldom available for the study of childhood disorders and cannot, of course, contribute to the study of normal function in healthy subjects. Fortunately, modern technology has produced new means to study the living brain.

Neuroimaging Techniques

Major technological advances over the past decade have facilitated brain imaging of intact human subjects. These include computerized axial tomography (CAT scan—Baker et al., 1975), nuclear magnetic resonance (NMR imaging—Kramer et al., 1981; James et al., 1982; Brownell et al., 1982), positron emission tomography (PET scan—Greenberg et al., 1981; Yarowsky and Ingvar, 1981), regional cerebral blood-flow (RCBF) studies (Larsen, 1978; Lassen, 1978; Yarowsky and Ingvar, 1981; Roland et al., 1981; Olesen et al.,

1982), and brain electrical activity mapping (BEAM — Duffy et al., 1979; Duffy, 1982). Although this chapter will primarily detail the last-named methodology, an understanding of the other techniques puts BEAM in perspective. Each method uses a computer to construct an image of the brain based on one or more measured parameters.

The CT and NMR scanners are used primarily for demonstration of anatomy. The CT scan provides a high-definition three-dimensional reconstruction of the brain's absorption of X-rays and has become the neurologist's standard tool for delineation of gross brain anatomy. The NMR display uses a high-density magnetic field to provide a similar three-dimensional reconstruction. As currently used, NMR images the concentration of hydrogen ions within the brain and thereby provides more information concerning differences in water content than the CT scan. For example, it may take days for anatomical change produced by a cerebrovascular accident to become visible by CT scan; the NMR delineates it much sooner, because of fluid shifts within infarcted and ischemic brain. The images provided are of comparable resolution. In contrast to CT, NMR is much less invasive — there are no known detrimental effects from strong magnetic fields. Moreover, NMR has the theoretical potential of providing more functionally useful information if images can eventually be obtained of phosphorus or even more complex substances such as specific neurotransmitters.

The PET scan and RCBF studies are methods designed to map brain *function*. In PET scanning, analogues of such metabolic substrates as glucose are administered after labeling with positron-emitting atoms. For instance, intravenously administered 18F-deoxyglucose (18FDG) enters functionally active brain cells, but because it is incompletely metabolized, it is temporarily "stuck" in the cells it has entered. A reconstruction of the autoradioactivity produced by 18FDG also gives anatomical information by showing structures with increased metabolism resulting from functional activation. Resolution of the PET scan image is less than that of the CT scan and NMR; unlike them, the PET is sensitive to cerebral function. Nonetheless, PET scanning is invasive and requires a nearby synchrotron to produce the short half-life positron-emitting compounds. PET has other disadvantages. It takes up to 20 min to complete preferential glucose uptake, and a comparable time for image construction by autoradiography. Thus, PET is primarily sensitive to functional state changes or clinical conditions that can be maintained for a substantial period of time and probably will not be able to delineate a fleeting thought or a brief judgment.

Many applications of RCBF imaging have consisted of mapping changes in blood flow induced by alterations in a subject's functional state (Yarowsky and Ingvar, 1981). Blood flow is believed to correspond closely to metabolic rate; hence localized changes in blood flow are taken to represent localized changes in brain metabolism induced by functional maneuvers. Unfortunately, the most detailed localizations are obtained by the intracarotid injection of labeled xenon (133Xe), making the procedure highly invasive (Roland et al., 1981). Less invasive routes of administration, such as inhalation of 133Xe, provide clinically useful information but are more prone to artifact and seldom used for physiological investigations (Olesen et al., 1982). Although PET, RCBF, and other functional mapping procedures have vastly augmented our understanding of cerebral specialization and asymmetry in the intact, living human, their invasive nature necessarily limits wide applicability, especially to normal subjects.

Brain Electrical Activity Mapping

There continues to be strong interest in localization of higher brain functions by study of brain electrical activity—the electroencephalogram (EEG—Klass and Daly, 1980) and sensory evoked potential (EP—Shagass, 1982). This interest stems from the fact that brain electrical activity is spontaneously present (EEG) or easily evoked (EP) and can be measured in a completely noninvasive, repeatable manner. EEG is of major clinical value for lesion localization, especially when the pathology produces a distinctive signature such as the spike and wave of epilepsy or the focal slow wave of cerebral abscess and tumor.

Unfortunately, EEG and EP studies have proved to be of limited use in the localization of higher functions. We believe that this does not stem from an inherent insensitivity of brain electrical activity to underlying brain function. Brain electrical activity, in fact, seems to contain not too little, but too much information to be easily interpreted by unaided visual inspection. Figure 5.1 shows a typical EEG and illustrates the problems inherent in visual inspection. As the electroencephalographer evaluates an EEG, he first searches for discontinuities such as the spike of epilepsy. Failing to find discontinuities, he turns to analysis of the EEG background activity. This process may be subdivided into four components:

(1) Spectral analysis. How much of each classic EEG frequency band (for instance, 8 to 10 Hz alpha) is there for each individual channel?

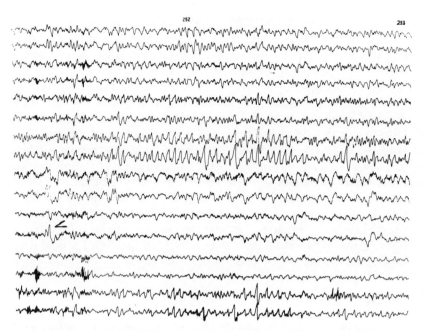

Fig. 5.1 *A 20-sec segment from a typical EEG. The arrow shows a discontinuity which stands out from the background and represents a clinically important paroxysm easily recognized by visual inspection. The remainder of this record is background; its evaluation visually is much more difficult.*

(2) Spatial mapping. How is the spectral information distributed across the scalp among electrodes?
(3) Temporal summation. How consistent or inconsistent are these activities over time?
(4) Statistical analyses. Are the results of the above analyses normal or abnormal?

The outcome of this complex process depends heavily on the skill of the electroencephalographer and is difficult to quantify.

Many disease states fail to produce obvious discontinuities, but manifest themselves as subtle changes of EEG background that frequently are hard to interpret. This is the probable explanation for the failure of EEG to have much effect on our understanding of diseases such as dyslexia, depression, schizophrenia, and cerebral palsy. Although abnormal EEGs are often reported, the findings are usually too nonspecific to provide clinically or physiologically useful information. Moreover, changes of higher brain function in normal subjects affect primarily the EEG background and are similarly difficult to detect.

The analysis of EP data by visual inspection is also quite complex. Sensory EPs are the brain's transient electrical response to external stimulation. To obtain an EP it is necessary to repeat the stimulation many times and perform signal averaging to eliminate the higher-voltage, ambient, resting-brain activity. Once signal averaging is complete, the resulting EP waveform is treated as if it were the response to a single stimulus presentation. The problem, illustrated in Fig. 5.2, is that EP morphology varies widely depending upon electrode placement. As with EEG, there is often too much information to be easily interpreted by unaided visual inspection.

In the mid-1970s our group started to develop a comprehensive, computerized procedure to aid the analysis of EEG and EP, so that the data could help in localization of both normal and abnormal brain function. We performed the requisite spectral, spatial, temporal, and statistical analyses by computer, thereby relieving the clinician/investigator of the necessity to perform those analyses in his own mind. The end result was a methodology now known as brain electrical activity mapping, or BEAM (Duffy et al., 1979; Duffy, 1982). The procedure is as follows.

In the case of EEG analysis, data from each of the 20 scalp electrodes are spectrally analyzed by means of the computerized Fast Fourier Transform (FFT) algorithm. Segments to be analyzed can range from 1 sec to 30 min. Thus, spectral analysis and temporal summation are automatically performed. Next, the amount of EEG energy in each of the classic EEG frequency bands is calculated from each electrode's spectral function. Finally, a topographic plot of the distribution of energy in each spectral range is created and displayed on a color graphics terminal; in other words, an automatic spatial analysis is performed.

For EP data, 20 waveforms are created for each stimulus modality, one from each of the 20 scalp electrodes. Each EP is analyzed for 512 msec following stimulation. There are 128 data points per curve, each representing 4 msec of time. A spatial analysis of the EP voltage across the head is carried out by topographic mapping (as done for EEG spectral energy). Each of the 128 images represents the spatial distribution of the EP during a 4-msec epoch. To allow visualization of changes in topographic distribution with time, the 128 images are viewed in rapid sequence on the color graphics terminal, a process known as cartooning. By topographic mapping and cartooning techniques, the important spatial and temporal characteristics of the EP are automatically summarized. Figure 5.2 details the topographic mapping process used in BEAM.

To facilitate the detection of abnormalities within a BEAM im-

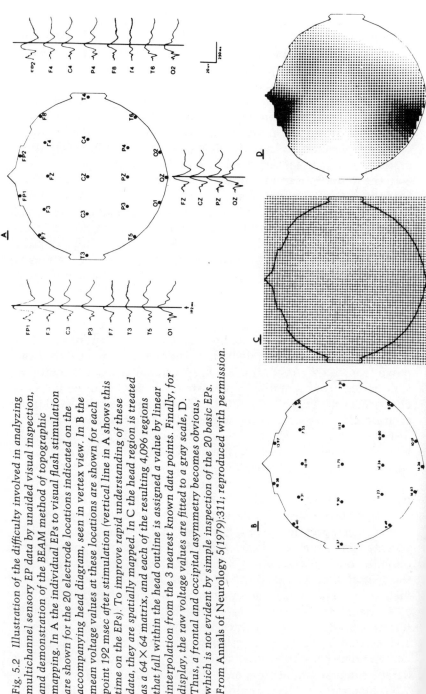

Fig. 5.2 Illustration of the difficulty involved in analyzing multichannel sensory EP data by unaided visual inspection, and demonstration of the BEAM method of topographic mapping. In A the individual EPs to visual flash stimulation are shown for the 20 electrode locations indicated on the accompanying head diagram, seen in vertex view. In B the mean voltage values at these locations are shown for each point 192 msec after stimulation (vertical line in A shows this time on the EPs). To improve rapid understanding of these data, they are spatially mapped. In C the head region is treated as a 64 × 64 matrix, and each of the resulting 4,096 regions that fall within the head outline is assigned a value by linear interpolation from the 3 nearest known data points. Finally, for display, the raw voltage values are fitted to a gray scale, D. Thus, a frontal and occipital asymmetry becomes obvious, which is not evident by simple inspection of the 20 basic EPs. From Annals of Neurology 5(1979):311; reproduced with permission.

age, a process known as significance probability mapping (SPM) was developed (Duffy et al., 1981). The data of a subject's image are automatically compared with those of a normalized control population. The original image is replaced, point by point, with a map of the degree of deviation from normal in units of standard deviation (the classic Z-statistic). Two groups may be similarly compared by means of a point-by-point calculation and display of Student's t-statistic. Thus, the Z-SPM topographic map delineates regions where one subject differs from a group, whereas the t-SPM displays where two groups differ from each other. SPM constitutes imaging of statistical properties contained within the original data; the details are procedurally outlined in Fig. 5.3. The result is a statistical delineation of abnormalities.

Let us look at three examples that illustrate the potential utility of BEAM in the localization of function and the determination of functional asymmetry.

NEUROPHYSIOLOGICAL STUDIES OF DYSLEXIA PURE For several years we have been studying children with specific reading disability, or dyslexia. This clinical entity was initially chosen because it represented one of the most readily definable functional abnormalities of the central nervous system not associated with consistent or diagnostic gross anatomical brain lesions by CT scan. Community surveys have shown that dyslexia affects 3.5% to 6.0% of school-age children (Rutter, 1978); moreover, 60% of all children referred to the Learning Disabilities Clinic at Children's Hospital, Boston, have a measurable degree of reading disability. Our goal was to see if BEAM could topographically locate regions where the brain electrical activity of dyslexics differed from that of nondyslexic controls.

To minimize variability in the clinical group, we chose to study only 9- to 11-year-old boys with established dyslexia. To further ensure homogeneity, we used the classification system of Hughes and Denckla (1978) and restricted the reading-disabled population to children who met the criterion for dyslexia pure. Such children must be of normal intelligence (full scale IQ of 95 to 115), without clinical neurological abnormality or emotional disturbance, and free of other forms of learning disability. Reading failure cannot be attributable of attentional deficit disorder (Conners, 1973) nor to lack of socioeconomic opportunity. The child must also have oral reading scores 1.5 years below expectancy (Rutter, 1978). The control group was matched for sex, age, socioeconomic status, and handedness but was at appropriate grade level on the Gray Oral

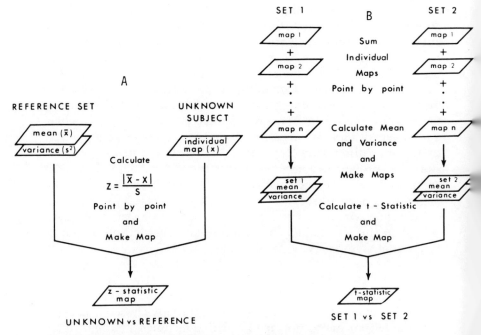

Fig. 5.3 Demonstration of significance probability mapping (SPM). In A, construction of a Z-SPM is diagrammed. An SPM image defines regions in which the individual subject differs statistically from the reference population. In B, construction of a t-SPM is diagrammed. The t-statistic map reveals regions in which the two populations differ significantly from each other. For display purposes an SPM image is formed representing the distribution of ranges of the value of t. From EEG and Clinical Neurophysiology 51 (1981):456; reproduced with permission.

Reading Test (Gray, 1963). Eight children meeting these stringent criteria were compared with 10 nondyslexic control boys. The results have been reported (Duffy et al., 1980a,b).

Spontaneous EEG was taken during 10 different testing conditions or states. These permitted recording during simple resting brain activity (eyes open or closed) and during tasks designed to activate the left hemisphere (speech and reading), the right hemisphere (music and geometric figures), and both hemispheres together (paired visual-verbal associations). The following were the 10 EEG test states:

(1) Speech (S): to listen carefully to a tape-recorded fairy tale ("The Elephant and the Butterfly," e.e. cummings) and answer simple questions about its content.

(2) Music (M): to listen to music ("Nutcracker Suite," Tchaikovsky).
(3) Kimura figures instruction (KFI): to memorize a set of 6 abstract geometric shapes presented by an examiner (Kimura, 1963).
(4) Kimura figures test (KFT): to select the 6 previously presented figures from a set of 38 figures, verbally indicating yes or no.
(5) Paired associates instruction (PAI): to learn a sound-symbol association between each of 4 abstract figures and a nonsense syllable spoken by the examiner, for example, "mog" (Vellutino, 1975).
(6) Paired associates test (PAT): to name each of the 4 abstract figures used in (5).
(7) Reading task instruction (RTI): to read silently 3 previously unread paragraphs (Gray, 1963).
(8) Reading task test (RTT): to identify whether 34 typed sentences presented by the examiner were included in the previous 3 paragraphs.
(9) Eyes open (EO): to relax but remain still with the eyes open.
(10) Eyes closed (EC): to relax but remain still with the eyes closed.
Approximately 3 min of data were recorded in a clinical EEG laboratory for each of the 10 states.

Brain activity and suitable trial markers were tape-recorded for subsequent off-line formation of average EPs in 3 test states. Two (visual and auditory EP) were recorded with no instruction other than to remain alert. One ("tight-tyke" auditory EP) required a phonological discrimination. These 3 EP test states were as follows:
(1) Visual evoked potential (VEP): over 500 flashes from a Grass PS-2 strobe stimulator were presented at random interstimulus intervals always exceeding 1 sec; the unit was set at intensity 8 and placed 20 cm from the subject's closed eyes.
(2) Auditory evoked potential (AEP): over 500 clicks were presented via earphones at 92 db sound pressure level.
(3) "Tight-tyke" auditory evoked potential (TTAEP): over 250 presentations of the tape-recorded word "tight" were randomly presented, intermixed with a similar number of the word "tyke"; subjects were required to count the number of "tights" heard for half the presentation and the number of "tykes" for the remainder of the presentation.

To reduce the large amounts of data collected, we employed the t-SPM technique (Duffy et al., 1981) as shown in Fig. 5.3. For each particular state t-SPMs were formed between the two groups. We noted large regional intergroup differences overlying the left temporal-parietal region. These were produced primarily during the

AEP, TTAEP, and RTI states. Much to our surprise, however, large regional intergroup differences were also found in both medial frontal lobes during the PAI, PAT, and RTI states. A few individual dyslexic subjects demonstrated regional differences in this region during the TTAEP state, but the finding was not sufficiently consistent to produce a TTAEP group-specific effect.

Theories of dyslexia based upon traditional aphasiology implicate the posterior speech regions centered about Wernicke's area and extending into the adjacent regions of the left parietal and temporal lobes. Indeed, most neurophysiological studies of dyslexia have been limited to these regions. In agreement with other reports (Torres and Ayers, 1968; Hughes and Park, 1969; Connors, 1970; Hanley and Sklar, 1976; John et al., 1977; Symann-Louett et al., 1977), our study confirmed differences in the left posterior quadrant between dyslexia-pure boys and controls.

Additional areas of difference were found in both hemispheres, especially in the medial frontal lobes (Fig. 5.4A and B). This was surprising, since the medial frontal lobes at one time were not recognized as involved in the language process. However, RCBF studies on subjects about to undergo brain surgery have shown that the two medial frontal regions are prominently and consistently involved during language processing tasks. During silent reading, reading aloud, and speaking, the medial frontal lobes are activated to the same degree as the more traditional regions in and around Broca's area and Wernicke's area (Larsen, 1978; Lassen, 1978). In other words, the medial frontal regions that we have shown to differ electrophysiologically between dyslexic and control boys appear to be among the extensive regions shown by RCBF studies to be involved in normal reading and speech. Several studies on lesions of the medial frontal region or on stimulation of it at operation confirm its role in speech function (Petit-Dutaillis et al., 1954; Guidetti, 1957; Alajouanine et al., 1959; Penfield and Roberts, 1959; Masdeu, 1978). Dyslexia pure may represent dysfunction within the entire widely distributed brain system devoted to language processing.

There is a known clinical overlap and statistical association of dyslexia with the hyperactive syndrome (Clements and Peters, 1962) and with attentional deficits (Douglas, 1976). Indeed, the need to exclude boys with elements of this syndrome proved to be the major reason for the small number of dyslexia-pure subjects we were able to recruit. The association of dyslexia and the attentional deficit disorder may be on the basis of some yet unspecified frontal lobe dysfunction.

This study, using small numbers of carefully chosen subjects, demonstrates the potential value of BEAM and SPM in the development of pathophysiological-anatomical correlations in the intact human. Our current studies of reading disability are concentrated in two areas: (1) assessment of the value of BEAM in the prediction from preschool data of first-grade reading problems; and (2) the search for electrophysiological correlates of subtypes of dyslexia pure.

INDIVIDUAL RESPONSES TO MUSIC AND SPEECH We have been investigating the potential of BEAM for localization of cortical function within an individual subject. In our early study of dyslexia, t-SPMs were used to localize regional differences between groups of dyslexic boys and normal-reading controls in the same activation or resting state. In our attempt to define state-dependent variation within individual subjects, repeated recordings of a resting baseline state were performed and compared by t-SPM to multiple recordings of data sampled during an activation paradigm.

We studied 12 normal right-handed males ranging in age from 10 to 27. Each was studied in 3 states: EO, SPE, and MUS. At least 60 separate EEG segments were spectrally analyzed to form the baseline group for each subject. Similarly, 60 EEG segments from SPE or MUS formed the second group. Thus t-SPM (EO × SPE) demonstrated regional change induced by speech, and t-SPM (EO × MUS) showed change induced by music. We hoped to expand upon prior electrophysiological work, which had suggested left-hemispheric activation by speech and right-hemispheric activation by music (McAdam and Whitaker, 1971; Morrell and Salamy, 1971; Doyle et al., 1974; Butler and Glass, 1974; Dolce and Waldeier, 1974; Galambos et al., 1975; Friedman et al., 1975; Davis and Wada, 1977; Shucard et al., 1977; Chapman et al., 1978; Wogan et al., 1979; Fried et al., 1980; Grabow et al., 1980; Shucard et al., 1981; Leubuscher, 1981).

As we anticipated, the EEG frequency range demonstrating the greatest change with state was alpha (8 to 12 Hz). Figure 5.4C and D illustrates the findings. Four of the 12 subjects showed the expected hemispheric change with speech and music. A large change limited to the left hemisphere was induced by speech stimulation. Music stimulation induced change that was greatest in the right hemisphere, but involved the left as well. This finding could be taken to support the hypothesis that either hemisphere may be active in music appreciation. Experienced musicians are more likely to demonstrate left-hemisphere activation, whereas normal nonmusicians

show less hemispheric specialization for this task (McKee et al., 1973; Bever and Chiarello, 1974; Davidson and Schwartz, 1977).
The results for the remaining 8 subjects were much less clear. Three demonstrated change involving most of both hemispheres, 3 failed to demonstrate significant change, and 2 demonstrated change limited to the occipital regions bilaterally. Thus the "correct" localization was correct in only 4 of 12 subjects or 33.3%. A retrospective evaluation of data revealed a probable explanation for the unusual results. By its very nature, t-SPM demonstrates all intergroup difference, no matter what the origin. The change in state from EO to SPE or MUS caused more central nervous system change than mere activation of specific processing circuitry, a problem noted by others (Gevins et al., 1979). Indeed, some of the subjects became bored and inattentive during EO, but were aroused and attentive during SPE and MUS. EEG background is very sensitive to changes in level of arousal and/or attention, probably more so than to subtle changes of the higher cortical functions. Accordingly, large changes in attention made the more subtle changes in brain function difficult to detect by t-SPM. To pursue the localization of function within an individual subject, we are currently developing new paradigms in which attention can be more readily equalized across states for all subjects.

INTRASUBJECT LOCALIZATION OF RHYTHM DISCRIMINATION Roland et al. (1981) performed RCBF studies to study focal activation of the human cerebral cortex during auditory discrimination. This group used the Seashore Rhythm Test (Saetveit et al., 1940) designed originally to identify musical talent and generally considered a right-hemisphere activating task. It comprises a two-alternative forced-choice discrimination of tone sequences, as illustrated in Fig. 5.5. Subjects are presented with two sequential tone patterns and asked whether they are the same or different.

The Roland group examined patients in whom intracarotid catheters had been placed for arteriographic purposes. Active epileptics were excluded. RCBF was measured by the clearance of 133Xe injected into the internal carotid artery on one side. The resulting data represented regional cortical energy consumption during the 40 sec necessary to measure the blood flow. Approximately 7 rhythm pairs were presented; responses were requested only outside of the measurement period. Results for a group of 12 subjects are summarized in Fig. 5.6. Note the activation in the right hemisphere involving posterior parietal and adjacent posterior temporal cortex and, in

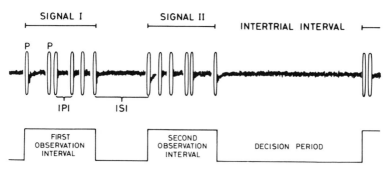

Fig 5.5 The sequence of events in a single presentation of the Seashore Rhythm Test. Two trains of tone pips are presented over a 3- to 4-second epoch: Signal I (first rhythm stimulus) and Signal II (second rhythm stimulus). Immediately afterward, the subject is asked if the two rhythms were the same or different. Figure from Roland et al., Journal of Neurophysiology 45(1981):1141; reproduced with permission.

addition, both medial and lateral frontal cortex. Because a lapse of time is necessary for blood-flow calculation, it was not possible to determine the time relations between the different regional activations during the observation and decision periods. The investigators suggest that the rhythm analysis depends on the areas activated on the right side.

We also studied Seashore Rhythm performance by means of BEAM. Our goals were threefold: (1) to see whether electrical recording would replicate the functional localization demonstrated by RCBF; (2) to delineate sequential activation, using the ability of BEAM to sample and study short EEG segments; and (3) to control carefully and/or delineate the effects of attention.

In our study of one right-handed 20-year-old male subject, the baseline was the EO state with a few additions. The subject was asked to fixate on a cross hair so as to control eye movement. Muscle artifact was eliminated by careful positioning of the body. The subject was told that at some undisclosed instant in the 6-min data collection period he might hear, see, or feel something unusual, and if so, should remember what he experienced. No stimulus was in fact presented, but the instruction served to promote attention and alertness. Next, the Seashore Rhythm task was explained to the subject. He was told that he would be periodically (randomly) interrupted and asked if the immediately preceding rhythm pairs matched. Eye fixation and muscle relaxation were maintained as for the EP state.

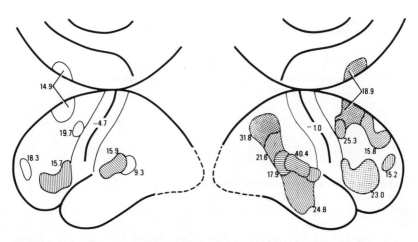

Fig. 5.6 Regional brain activation produced by the Seashore Rhythm Test, measured by regional cerebral blood-flow measurement, as reported by Roland. Note that the right hemisphere is more activated than the left. Over the right hemisphere there are two broad regional activations, in the frontal lobe and the posterior temporal-parietal region. Figure from Roland et al., Journal of Neurophysiology 45(1981):1144; reproduced with permission.

Data from the 20 scalp electrodes were tape-recorded for later off-line spectral analysis. An event marker indicating the end of the second rhythm sequence was also recorded on tape. This marker signaled the end of stimulus presentation and the beginning of the period used for stimulus evaluation — the decision-making period. Data were then sampled for periods prior to and immediately following the trial marker, allowing separate evaluation of stimulus presentation (S) and decision (D) phases. Two seconds of each period were recorded for each presentation. Approximately 60 artifact-free 2-sec segments were obtained for both S and D states. We formed t-SPMs comparing the EO and S states, in order to delineate regional cortical change during stimulus presentation. Similarly the t-SPM from the EO and D comparison delineated regions showing differences during the decision-making process.

The results are shown in Fig. 5.7. For the alpha frequency range (8 to 12 Hz), topographic differences were evident for the S and D periods by t-SPM. The EO × S comparison (Fig. 5.7A) revealed major changes in the right and left posterior temporal and occipital regions, and lesser changes in prefrontal areas. The EO × D comparison revealed a different pattern, with maximal change in the right frontal region and much less change in the right posterior temporal region (Fig. 5.7B). Examination of higher frequency ranges provided better definition of the frontal lobe change and poorer definition of

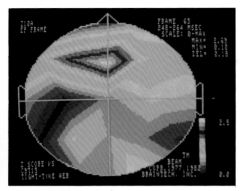

A. TTAEP 248–264 msec; Z_{max} = 2.69

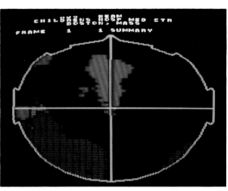

B. Summary t-SPM

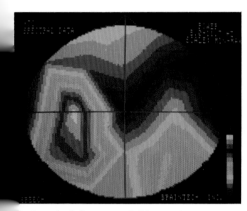

C. Speech alpha; t_{max} = 2.74

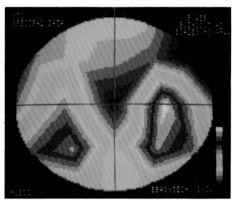

D. Music alpha; t_{max} = 3.39

Fig. 5.4 Each of these 4 BEAM images shows data within a schematic map of the head seen from vertex view, nose above, right ear to the right, and so on. The Z-SPM of a single 10-year-old dyslexic boy is shown in A. Note the unexpected bilateral medial frontal abnormality, which peaked (white area) at 2.69 standard deviations from normal. B, adapted from an earlier BEAM device, summarizes the results of our study of dyslexic boys. Each colored region demarcates the areas where the dyslexics, as a group, differed from the controls, as a group. This composite t-SPM shows only regions with $p < 0.02$ level (2-tailed). In addition to the involvement of the classic left posterior language-related region, bilateral medial frontal group difference is also noted. C and D are t-SPMs used to delineate regions activated during speech and music stimulation of one subject. In C, left hemispheric function is elicited by listening to a sophisticated fairy tale. In D, bihemispheric—mostly right-sided—functional activation is produced by listening to classical music.

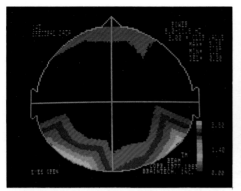

A. Alpha EO × S; t_{max} = 3.18

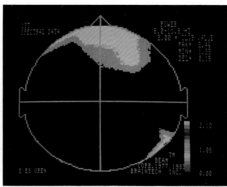

B. Alpha EO × D; t_{max} = 2.46

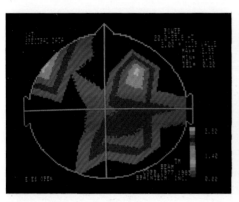

C. Beta-3 EO × S; t_{max} = 2.93

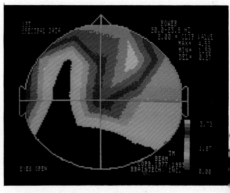

D. Beta-3 EO × D; t_{max} = 4.52

Fig. 5.7 Regional changes of brain electrical activity in one subject induced by the Seashore Rhythm Test, as shown by measurements of t-SPM. In A and B, changes are shown for the alpha (8 to 12 Hz) EEG frequency range; in C and D, for beta 3 (20 to 24 Hz); and in E, for delta (0 to 4 Hz). T-SPM in A and C compare the baseline (EO) state with the stimulus presentation (S) state. T-SPM in B, D, and E compare EO with the decision-making period, D. Note that for alpha and beta more posterior differences are seen during the S period, more anterior differences during the D period, and more change is seen over the right hemisphere. For delta, change is global and limited to the D state. Above each image are shown the frequency range, comparison made, and maximum t value.

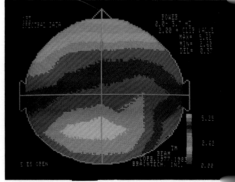

E. Delta EO × D; t_{max} = 5.62

the posterior change. For the 20 to 24 Hz range some posterior and left frontal change was noted by EO × S (Fig. 5.7C) in addition to the prominent right frontal change. By EO × D (Fig. 5.7D), the change was almost exclusively in the right frontal area and of greater magnitude than seen for EO × S.

The most surprising finding was evident within the delta frequency range (0 to 4 Hz). The EO × S comparison demonstrated no significant change. However, the EO × D comparison showed very significant and broadly symmetrical change stemming from a selective reduction of delta during the decision epoch (Fig. 5.7E). Delta is generally believed to mirror such factors as level of consciousness and degree of attention. We suggest that the profound and global reduction of delta during the decision process represents a general increase in cerebral activation (hence delta reduction) during the difficult task of mentally comparing two rhythm sequences. Evidently this change of low-frequency brain electrical activity is either not associated with changes of blood flow, or else the change is of such short duration that it cannot be appreciated by current RCBF techniques.

By mapping the change in brain electrical activity of a single subject during rhythm discrimination, we have confirmed the separate posterior and anterior regional activations reported by Roland et al. (1981). Moreover, we suggest that during rhythm presentation the posterior regions show the greater change, whereas during the decision period the frontal areas are more active. The posterior regions may be primarily involved in the reception, recall, and initial analysis of complex auditory stimuli, the frontal region primarily in the evaluation and decision-making process. The latter also appears to require participation of a more global activating mechanism.

Although topographic mapping of scalp-recorded brain electrical activity is now performed in a growing number of neurophysiological laboratories, it is often stated that the field is in its infancy and its ultimate contribution to our understanding of the brain in health and disease is not now predictable. Yet the value of topographic mapping of EEG was recognized in the 1940s and 1950s. Most of the early techniques employed multichannel cathode-ray or glow-tube displays that could be observed on-line or photographed on movie film for viewing at slower speed (Goldman et al., 1948, 1949; Lilly, 1950; Walter and Shipton, 1951; Petsche and Marko, 1954; Shipton, 1956, 1963; Livanov et al., 1956; and Walter, 1957).

The availability of the digital computer in the 1960s facilitated the generation of topographic maps that had previously been drawn laboriously by hand. Many methods have been employed to represent scalp activity, including spatiotemporal mapping (Rémond, 1955; Rémond et al., 1969; Lesevre and Rémond, 1970; Rémond and Offner, 1978), isopotential contour-line drawings (Lehmann, 1971; Goff et al., 1977; Ragot and Rémond, 1978; Vaughn and Ritter, 1979), three-dimensional representations (Harris et al., 1969; Childers et al., 1972; Halliday et al., 1972), gray scale numeric or dot density plots (Estrin and Uzgalis, 1969; Ueno et al., 1975; Dubinsky and Barlow, 1980; Buchsbaum et al., 1982), and colored gray scale or pseudocolor plots (Duffy et al., 1979; Nagata et al., 1980, 1981a,b).

It was not until the middle to late 1970s that these methods were widely applied to neurological patients (Ueno et al., 1975; Duffy et al., 1979). Two of the most sophisticated display systems were implemented by Harris et al. (1969) and by Estrin and Uzgalis (1969). Both involved computer plotting of electrophysiological data. It is surprising that many more recent articles have used much simpler display systems, including hand-drawn plots. In a review article Petsche (1976) suggested features to be incorporated into a topographic display system of the future. Many of these elements had, however, been implemented previously by the Harris group (1969).

The reasons for the lack of application of technology already developed are perhaps twofold. First, the early systems may have been operationally difficult, inconvenient, or expensive to use for anything more than a circumscribed research project. Second, the better systems appear to have been developed by engineers, and the lack of application to clinical problems may have reflected a communications gap between engineers and physicians. In the 1980s, however, the ubiquitous nature of the computer, including applications within the home, has tended to eliminate such problems. Many laboratories have begun to apply modern computer technology to studies of brain electrical activity. Recent work has explored the role of topographic mapping in localization of anatomical landmarks (Duff, 1980), the pathophysiology of dyslexia (Duffy et al., 1980a,b), cerebrovascular disease (Nagata and Mizukami, 1981; Nagata et al., 1982), basic auditory physiology (Wood and Walpow, 1982), the topographic signature of epilepsy (Lombroso and Duffy, 1980), the localization of cerebral processing (Gevins et al., 1983), and the pathophysiology of schizophrenia (Morihisa et al., 1983; Morstyn et al., 1983).

There seems to be every reason to expect that topographic display techniques will continue to expand the value of electrophysiological data and will complement information obtained from other brain imaging techniques. Such methodologies should permit useful clinical correlations and thereby expand our knowledge of cerebral specialization and functional asymmetry.

References

Alajouanine, T., Castaigne, P., Sabouraud, O., and Contamin, F. 1959. Palialie paroxystique et vocalisations itératives au cours de crises épileptiques par lésion intéressant l'aire motrice supplémentaire. *Rev. Neurol. (Paris)* 101:685–697.
Baker, H. L., Houser, O. W., Campbell, J. K., and others. 1975. Computerized tomography of the head. *J. Am. Med. Assoc.* 233:1304–8.
Bever, T. G., and Chiarello, R. J. 1974. Cerebral dominance in musicians and nonmusicians. *Science* 185:537–539.
Brownell, G. L., Budinger, T. F., Lauterbur, P. C., and McGeer, P. L. 1982. Positron tomography and nuclear magnetic resonance imaging. *Science* 215:619–626.
Buchsbaum, M. S., Rigal, F., Coppola, R., and others. 1982. A new system for gray level surface distribution maps of electrical activity. *Electroenceph. Clin. Neurophys.* 53:237–242.
Butler, S. R., and Glass, A. 1974. Asymmetries in the electroencephalogram associated with cerebral dominance. *Electroenceph. Clin. Neurophys.* 36:481–491.
Chapman, R. M., McCrary, J. W., Chapman, J. A., and Bragdon, H. R. 1978. Brain responses related to semantic meaning. *Brain and Lang.* 5:195–204.
Childers, D. G., Perry, N. W., Jr., Halpeny, O. S., and others. 1972. Spatiotemporal measures of cortical functioning in normal and abnormal vision. *Comp. Biomed. Res.* 5:114–130.
Clements, S. D., and Peters, J. E. 1962. Minimal brain dysfunction in the school-age child. *Arch. Gen. Psychiat.* 6:185–197.
Connors, C. K. 1970. Cortical visual evoked response in children with learning disorders. *Psychophysiology* 7:418–428.
Connors, C. K. 1973. Psychological assessment of children with minimal brain dysfunction. *Ann. N.Y. Acad. Sci.* 205:283–302.
Davidson, R. J., and Schwartz, G. E. 1977. The influence of musical training on patterns of EEG asymmetry during musical and non-musical self-generation tasks. *Psychophysiology* 14:58–63.
Davis, A. E., and Wada, J. A. 1977. Lateralization of speech dominance by spectral analysis of evoked potentials. *J. Neurol. Neurosurg. Psychiat.* 40:1–4.
Dolce, G., and Waldeier, H. 1974. Spectral and multivariate analysis of EEG changes during mental activity in man. *Electroenceph. Clin. Neurophys.* 36:577–584.
Douglas, V. 1976. Perceptual and cognitive factors as determinants of learning disabilities: a review chapter with special emphasis on atten-

tional factors. In R. M. Knights and D. J. Bakker, eds., *The Neuropsychology of Learning Disorders.* Baltimore: University Park Press.
Doyle, J. C., Ornstein, R., and Galin, D. 1974. Lateral specialization of cognitive mode. II. EEG frequency analysis. *Psychophysiology* 11:567-578.
Drake, W. E. 1968. Clinical and pathological findings in a child with a developmental learning disability. *J. Learning Dis.* 1:9-25.
Dubinsky, J., and Barlow, J. S. 1980. A simple dot-density topogram for EEG. *Electroenceph. Clin. Neurophys.* 48:473-477.
Duff, T. A. 1980. Topography of scalp recorded potentials evoked by stimulation of the digits. *Electroenceph. Clin. Neurophys.* 49:452-460.
Duffy, F. H. 1982. Topographic display of evoked potentials: clinical applications of brain electrical activity mapping (BEAM). *Ann. N.Y. Acad. Sci.* 388:183-196.
Duffy, F. H., Burchfiel, J. L., and Lombroso, C. T. 1979. Brain electrical activity mapping (BEAM): a method for extending the clinical utility of EEG and evoked potential data. *Ann. Neurol.* 5:309-321.
Duffy, F. H., Denckla, M. B., and Sandini, G. 1980a. Dyslexia: regional differences in brain electrical activity by topographic mapping. *Ann. Neurol.* 7:414-420.
Duffy, F. H., Denckla, M. B., Bartels, P. H., and others. 1980b. Dyslexia: automated diagnosis by computerized classification of brain electrical activity. *Ann. Neurol.* 7:421-428.
Duffy, F. H., Bartels, P. H., and Burchfiel, J. L. 1981. Significance probability mapping: an aid in the topographic analysis of brain electrical activity. *Electroenceph. Clin. Neurophys.* 51:455-462.
Estrin, T., and Uzgalis, R. 1969. Computerized display of spatio-temporal EEG patterns. *IEEE Trans. Bio-Med. Eng.* BME-16:192-196.
Fried, I., Ojemann, G., and Fetz, E. E. 1981. Language-related potentials specific to human language cortex. *Science* 212:353-356.
Friedman, D., Simson, R., Ritter, W., and Rapin, I. 1975. Cortical evoked potentials elicited by real speech words and human sounds. *Electroenceph. Clin. Neurophys.* 38:13-19.
Galaburda, A. M., and Kemper, T. L. 1979. Cytoarchitectonic abnormalities in developmental dyslexia. *Ann. Neurol.* 6:94-100.
Galambos, R., Benson, P., Smith, T. S., and others. 1975. On hemispheric differences in evoked potentials to speech stimuli. *Electroenceph. Clin. Neurophys.* 39:279-283.
Gevins, A. S., Zeitlin, G. M., Doyle, J. C., and others. 1979. EEG patterns during 'cognitive' tasks. II. Analysis of controlled tasks. *Electroenceph. Clin. Neurophys.* 47:704-710.
Gevins, A. S., Doyle, J. C., Cutillo, B. A., and others. 1981. Electrical potentials in human brain during cognition: new method reveals dynamic patterns of correlation. *Science* 213:918-922.
Gevins, A. S., Schaffer, R. E., Doyle, J. C., and others. 1983. Shadows of thought: shifting lateralization of human brain electrical patterns during brief visuomotor task. *Science* 220:97-99.

Goff, G. D., Matsumiya, Y., Allison, T., and Goff, W. R. 1977. The scalp topography of human somatosensory and auditory evoked potentials. *Electroenceph. Clin. Neurophys.* 42:57–76.

Goldman, S., Vivian, W. E., Chien, C. K., and Bowes, N. H. 1948. Electronic mapping of the activity of the heart and brain. *Science* 108:720–723.

Goldman, S., Santleman, N. W. F., Vivian, W. E., and Goldman, D. 1949. Traveling waves in the brain. *Science* 109:524.

Grabow, J. D., Aronson, A. E., Offord, K. P., and others. 1980. Hemispheric potentials evoked by speech sounds during discrimination tasks. *Electroenceph. Clin. Neurophys.* 49:48–58.

Gray, W. S. 1963. *Gray Oral Reading Test.* Indianapolis: Bobbs-Merrill.

Greenberg, J. H., Reivich, M., Alavi, A., and others. 1981. Metabolic mapping of functional activity in human subjects with the 18F fluorodeoxyglucose technique. *Science* 212:678–680.

Guidetti, B. 1957. Désordres de la parole associés à des lésions de la surface interhémisphérique frontale postérieure. *Rev. Neurol. (Paris)* 97:121–131.

Halliday, A. M., McDonald, W. I., and Mushin, J. 1972. Delayed visual evoked response in optic neuritis. *Lancet* 1:982–985.

Hanley, J., and Sklar, B. 1976. Electroencephalographic correlates of developmental reading disorders: computer analyses of recordings from normal and dyslexic children. In G. Leisman, ed., *Basic Visual Processes and Learning Disability.* Springfield, Illinois: Charles C Thomas, pp. 217–243.

Harris, J. A., Melry, G. M., and Bickford, R. M. 1969. Computer-controlled multidimensional display device for investigation and modeling of physiologic systems. *Comp. Biomed. Res.* 2:519–536.

Hughes, J. R., and Denckla, M. B. 1978. Outline of a pilot study of electroencephalographic correlates of dyslexia. In A. L. Benton and D. Pearl, eds., *Dyslexia: An Appraisal of Current Knowledge.* New York: Oxford University Press.

Hughes, J. R., and Park, G. E. 1969. Electro-clinical correlations in dyslexic children. *Electroenceph. Clin. Neurophys.* 26:119.

James, A. E. Jr., Price, R. R., Rollo, D., and others. 1982. Nuclear magnetic resonance imaging. A promising technique. *J. Am. Med. Assoc.* 247:1331–34.

John, E. R., Karmel, B. Z., and Corning, W. C., and others. 1977. Neurometrics. *Science* 196:1393–1410.

Kimura, D. 1963. Right temporal lobe damage. Perception of unfamiliar stimuli after damage. *Arch. Neurol.* 8:264–271.

Klass, D. W., and Daly, D. D., eds. 1980. *Current Practice of Clinical Electroencephalography.* New York: Raven Press.

Kramer, D. M., Schneider, J. S., Rudin, A. M., and Lauterbur, P. C. 1981. True three-dimensional nuclear magnetic resonance zeugmatographic images of a human brain. *Neuroradiology* 21:239–244.

Larsen, B., Skinhoj, E., and Lassen, N. A. 1978. Variations in regional corti-

cal blood flow in the right and left hemispheres during automatic speech. *Brain* 101:193-209.
Lassen, N. A., Ingvar, D. H., and Skinhoj, E. 1978. Brain function and blood flow. *Sci. Am.* 239:62-71.
Lehmann, D. 1971. Multichannel topography of human alpha EEG fields. *Electroenceph. Clin. Neurophys.* 31:439-449.
Lesevre, N., and Rémond, A. 1970. Influence des contrastes sur les réponses évoquées visuelles. *Rev. Neurol. (Paris)* 122:505-516.
Leubuscher, H. J. 1981. Dependence of EEG spectral power density on the meaning of stimuli. *Activ. Nerv. Sup.* 23:92-96.
Lilly, J. C. 1950. A method of recording the moving electrical potential gradients in the brain: the 25 channel bantron and electroiconograms. In *Conference on Electronics in Nucleonics and Medicine.* New York: American Institute of Electrical Engineers, pp. 37-43.
Livanov, M. N., Ananiev, V. M., and Bekhtereva, N. P. 1956. Electroencephaloscopic studies on bio-electric map of the cerebral cortex in cerebral tumors and injuries. *Zh. Nevropat. Psikhiat.* 56:788-790.
Lombroso, C. T., and Duffy, F. H. 1980. Brain electrical activity mapping as an adjunct to CT. In R. Canger and J. K. Penry, eds., *Advances in Epileptology: Proceedings of the XI Epilepsy International Symposium.* New York: Raven Press, pp. 83-88.
McAdam, D., and Whitaker, H. A. 1971. Language production: electroencephalographic localization in the normal human brain. *Science* 172:499-502.
McKee, G., Humphrey, B., and McAdam, D. W. 1973. Scaled lateralization of alpha activity during linguistic and musical tasks. *Psychophysiology* 10:441-443.
Masdeu, J. C., Schoene, W. C., and Funkenstein, H. 1978. Aphasia following infarction of the left supplementary motor area: a clinicopathologic study. *Neurology (Minn.)* 28:1220-23.
Morihisa, J. M., Duffy, F. H., and Wyatt, R. J. 1983. Brain electrical activity mapping (BEAM) in schizophrenic patients. *Arch. Gen. Psychiat.* 40:719-728.
Morrell, L. K., and Salamy, J. G. 1971. Hemispheric asymmetry of electrocortical responses to speech stimuli. *Science* 174:164-166.
Morstyn, R., Duffy, F. H., and McCarley, R. W. 1983. Altered P300 topography in schizophrenia. *Arch. Gen. Psychiat.* 40:729-734.
Nagata, K., and Mizukami, M. 1981. Computed mapping of EEG/evoked potential (CME). *Image Info.* 13:587-595. (In Japanese.)
Nagata, K., Araki, G., Mizukami, M., and others. 1980. Computed mapping electroencephalogram (CME) in cerebral infarction—a comparative study with CT and regional blood flow study. *No To Shinke* 32:1149-57. (English abstract; author's translation.)
Nagata, K., Tagama, T., and Mizukami, M. 1981. Study of computed mapping of EEG following extra-intracranial bypass operation. *Clin. Electroenceph. (Osaka)* 25:19-28. (In Japanese.)
Nagata, K., Mizukami, M., Araki, G., and others. 1982. Topographic elec-

troencephalographic study of cerebral infarction using computed mapping of the EEG. *J. Cereb. Blood Flow Metab.* 2:79-88.

Olesen, J., Lauritzen, M., Tfelt-Hansen, P., and others. 1982. Spreading cerebral oligemia in classical and normal cerebral blood flow in common migraine. *Headache* 22:242-248.

Penfield, W., and Roberts, L. 1959. *Speech and Brain Mechanisms*. Princeton: Princeton University Press.

Petit-Dutaillis, D., Guiot, G., Messimy, R., and Bourdillon, C. H. 1954. A propos d'une aphémie par atteinte de la zone motrice supplémentaire de Penfield, au cours de l'évolution d'un anévrisme artério-véneux. Guérison de l'aphémie par ablation de la lésion. *Rev. Neurol. (Paris)* 90:95-106.

Petsche, H. 1976. Topography of the EEG: survey and prospects. *Clin. Neurol. Neurosurg.* 79:15-28.

Petsche, H., and Marko, A. 1954. Das Photozellentoposkop, eine einfache Methode zur Bestimmung der Feldverteilung und Ausbreitung hirnelektrischer Vorgänge. *Arch. Psychiat.* 192:447-458.

Ragot, R. A., and Rémond, A. 1978. EEG field mapping. *Electroenceph. Clin. Neurophys.* 45:417-421.

Rémond, A. 1955. Orientations et tendances des méthodes topographiques dans l'étude de l'activité éléctrique du cerveau. *Rev. Med. Liège* 10:299-302.

Rémond, A., Lesevre, N., Joseph, J. P., and others. 1969. The alpha average. I. Methodology and description. *Electroenceph. Clin. Neurophys.* 26:245-265.

Roland, P. E., Skinhoj, E., and Lassen, N. A. 1981. Focal activations of human cerebral cortex during auditory discrimination. *J. Neurophys.* 45:1139-51.

Rutter, M. 1978. Prevalence and types of dyslexia. In A. L. Benton and D. Pearl, eds., *Dyslexia—An Appraisal of Current Knowledge*. New York: Oxford University Press.

Saetveit, J. G., Lewis, D., and Seashore, C. G. 1940. Revision of the Seashore measures of musical talent. *Univ. Iowa Study, Aims Prog. Res. No. 65*. Iowa City: University of Iowa Press.

Shagass, C. 1982. *Evoked Potentials in Psychiatry*. New York: Plenum Press.

Shipton, H. W. 1956. A new electrotoposcope using a helical scan. *Proc. Electrophys. Technolog. Assoc.* 2:2-11.

Shipton, H. W. 1963. A new frequency selective toposcope for electroencephalography. *Med. Electronic Biol. Eng.* 1:483-496.

Shucard, D. W., Shucard, J. L., and Thomas, D. G. 1977. Auditory evoked potentials as probes of hemispheric differences in cognitive processing. *Science* 197:1295-98.

Shucard, D. W., Cummins, K. R., Thomas, D. G., and Shucard, J. L. 1981. Evoked potentials to auditory probes as indices of cerebral specialization of function—replication and extension. *Electroenceph. Clin. Neurophys.* 52:389-393.

Symann-Louett, N., Gascon, G. G., Matsumiya, Y., and others. 1977. Wave form difference in visual evoked responses between normal and reading disabled children. *Neurology (Minn.)* 27:156-159.

Torres, R., and Ayers, F. W. 1968. Evaluation of the electroencephalogram of dyslexic children. *Electroenceph. Clin. Neurophys.* 24:281-294.

Ueno, S., Matsuoka, S., Mizoguchi, T., and others. 1975. Topographic computer display of abnormal EEG activities in patients with CNS diseases. *Mem. Fac. Eng., Kyushu Univ.* 34:195-209.

Vaughn, H. G., Jr., and Ritter, W. 1970. The sources of auditory evoked responses recorded from the human scalp. *Electroenceph. Clin. Neurophys.* 28:360-378.

Vellutino, F. R., Steger, J. A., Harding, C. J., and others. 1975. Verbal versus nonverbal paired-associates learning in poor and normal readers. *Neuropsychologia* 13:75-82.

Walter, W. G. 1957. The brain as a machine. *Proc. Roy. Soc. Med.* 50:799-808.

Walter, W. G., and Shipton, H. W. 1951. A new toposcopic display system. *Electroenceph. Clin. Neurophysiol.* 3:281-292.

Wogan, M., Kaplan, C. D., Moore, S. F., and Epro, R. 1979. Sex difference and task effects in lateralization of EEG-alpha. *Internat. J. Neurosci.* 8:219-223.

Wood, C. C., and Walpow, J. R. 1982. Scalp distribution of human auditory evoked potentials. II. Evidence for overlapping sources and involvement of auditory cortex. *Electroenceph. Clin. Neurophys.* 54:25-38.

Yarowsky, P. J., and Ingvar, D. H. 1981. Symposium summary. Neuronal activity and energy metabolism. *Fed. Proc.* 40:2353-62.

Chapter 6
Asymmetrical Lesions in Dyslexia

Thomas L. Kemper

A structural basis for developmental dyslexia has been suspected from the time of the first case report by W. Pringle Morgan (1896). Morgan's first case was a 14-year-old boy who, although seemingly intelligent, was incapable of learning to read even with special tutoring. Morgan believed that he had word blindness, but not letter blindness, and postulated a lesion in the left angular gyrus. The previous year James Hinshelwood (1895), an eye surgeon at the Glasgow Eye Infirmary, had written a paper on acquired word blindness. Morgan corresponded with Hinshelwood, indicating that it was this paper that had prompted him to publish his report (Critchley, 1964). Hinshelwood subsequently became the authority on congenital word blindness, publishing numerous papers that culminated in a monograph in 1917. In agreement with Morgan, he also postulated the presence of a lesion in the left angular gyrus in congenital word blindness.

The literature then remained almost silent regarding the possible presence of a lesion as the underlying cause of the disorder. A vast amount of data was accumulated about developmental dyslexia, including the frequent presence of a positive family history, prevalence among males, and frequent association with left-handedness. A prominent investigator in the period between the two World Wars, Samuel T. Orton, felt that no structural lesion was present. He viewed the brain as symmetrical, with a rivalry between equipotential areas on opposite sides in the two hemispheres; the symptoms he regarded as the result of lack of dominance of one area over the other (Orton, 1925). Despite his extensive training in both brain pathology and cytoarchitectonics, he himself never studied the brain of a dyslexic individual.

Although a description by William Drake (1968) of malformations in the brain of a well-documented case of developmental dyslexia was generally neglected at the time, strong evidence for a structural basis has recently become available. There have now been three autopsy studies, all showing malformations of the brain. Cranial CAT scan studies have been published describing deviations from the normal asymmetry of the two cerebral hemispheres (Hier et al., 1978; Rosenberger and Hier, 1980; Haslam et al., 1981), but these findings have been disputed and an EEG-evoked response study (Duffy et al., 1980) has shown abnormalities confined predominantly to the left cerebral hemisphere. In this chapter the autopsy cases will be reviewed and discussed.

Case Reports

CASE 1 The case reported by Drake (1968) was that of a 14-year-old boy who died of a brainstem hemorrhage secondary to a vascular malformation. No abnormalities had been present in pregnancy, birth, or early development. In school, although promoted each year, the child had progressive difficulty with reading, particularly in the fifth and sixth grades. Testing in the sixth grade showed that reading skill was 3 years below average. His difficulty was most marked when "words had to be read in the context of sentences or paragraphs when sequence and directional factors are important." His IQ on the California Test of Mental Maturity was 120 in the first grade, but only 86 and 87 in the fourth and fifth grades. Full-scale IQ scores were 101 and 96 on the Wechsler Intelligence Scale for Children (WISC) in the fifth and sixth grades. The respective verbal IQs were 99 and 96, and performance IQs were 103 and 97. Although predominantly right-handed, he was observed to use his left hand frequently in many tasks. In the family history were mixed dominance, "visual and learning problems," migraine headaches, and vascular malformations. His only brother was described as having a serious reading problem.

Many abnormalities were noted in a brief description of the brain. These included abnormal gyri in the parietal regions with the "cortical pattern . . . disrupted by penetrating deep gyri that appeared disconnected" and thinning of related areas of the corpus callosum. Microscopic examination showed that the cortex was "more massive than normal, the lamination tended to be columnar, the nerve cells . . . spindle shaped, and there were numerous ectopic neurons in the white matter that were not collected into distinct heterotopias." The brain weight was not provided.

CASE 2 The brains of the other two cases were studied in gapless whole-brain serial histological sections. The findings on the first of these have been published (Galaburda and Kemper, 1979; Galaburda and Eidelberg, 1982). The subject died in an accident at age 20. There had been no abnormalities in pregnancy, delivery, or early development. In early childhood the boy was clumsier than his siblings, and speech in full sentences was delayed until after age 3. Difficulties with reading and spelling were noted soon after he started elementary school, and he repeated the first grade, at which time a diagnosis of developmental dyslexia was made. His difficulties persisted despite special tutoring. At 19 years of age, he received scores for paragraph meaning at grade level 3.5, word meaning at 4.0, and spelling at 3.5 on the Stanford Achievement Test. On the Gray Oral Paragraphs and the Gates-MacGinitie reading test he scored at the 4.0 grade level. At 6 years of age IQ was 105 on the Stanford-Binet, and at 13 years he scored a full-scale IQ of 88, performance IQ of 83, and verbal IQ of 95 on the WISC.

Neurological evaluation at 18 years of age showed mild difficulties with right-left orientation and finger recognition. A dichotic listening test showed marked right-ear superiority. Starting at age 16, he had nocturnal seizures that were easily controlled with phenytoin. Routine electroencephalograms were normal, except for one sleep study that showed borderline slowing over the right cerebral hemisphere.

The patient was left-handed, as were several other members of the family. He was the youngest of 4 siblings, 3 males and 1 female. Both brothers and the father were slow readers, but not the mother or the sister.

At the time of autopsy the brain showed no gross abnormality and weighed 1,576 g after formalin fixation. Whole-brain serial sections showed a wider left than right cerebral hemisphere throughout. Reconstruction of the planum temporale in the cerebral hemispheres showed no asymmetry in size. However, the one on the left, which corresponds to part of Wernicke's speech area, showed an extensive area of abnormal polymicrogyric cortex (Fig. 6.1). Adjacent molecular layers were fused, and there were abnormal cytoarchitectonic features. In many areas the cortex was abnormally thin. A cell-free area in midcortical location, commonly found in this malformation, was not observed. Based on cytoarchitectonic parcellations of the auditory areas, this malformation was thought to be predominantly in area Tpt.

Closely associated with this polymicrogyric cortex were two other less striking cortical malformations: small focal accumula-

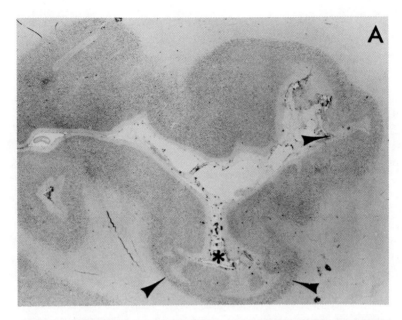

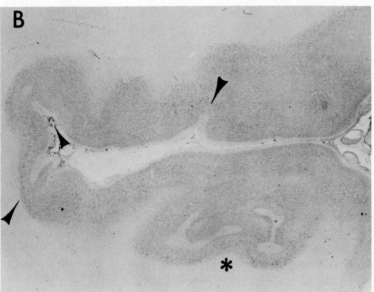

Fig. 6.1 Polymicrogyric cortex (*) at two levels in the serial sections of case 2. A shows the beginning of this malformation, and B its middle part. In the adjacent cortex are abnormal, fused microsulci, many with unusually thin cortex at the depths of the sulci (arrows).

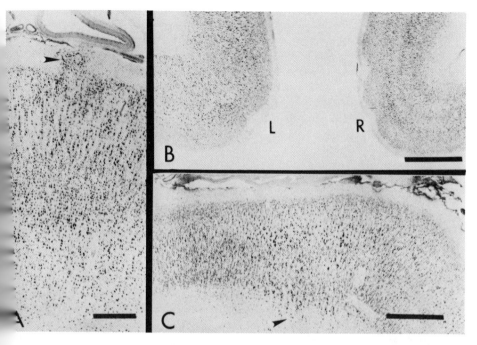

Fig. 6.2 *Minor malformations in case 2:* A, *a focal accumulation of neurons in layer I (arrow);* B, *the rostral cingulate cortex with extensive focal dysplasias on the left side (L);* C, *a single small area of focal dysplasia (arrow). Illustration from Galaburda and Kemper (1979); by permission of Little, Brown.*

tions of ectopic neurons in layer I (Fig. 6.2A), and scattered focal cortical dysplasias characterized by large neurons extending from the subcortical white matter to all neuronal cell layers but predominantly the deeper layers (Figs. 6.2B and C). Both types of lesion were confined to the left cerebral hemisphere. The focal accumulations of neurons in layer I varied in size from a few scattered neurons to wartlike projections with the upper cortical layers participating in the malformation (Fig. 6.3A). In the larger lesions a tuft of radially directed myelinated fibers was present in the center (Fig. 6.3B). Many lesions were associated with a prominent, centrally placed blood vessel. Unlike the focal cortical dysplasias, the ectopic collections showed a close topographic relation to the polymicrogyria (Fig. 6.4). When 8 fused sulci at the edge of the microgyric cortex were followed in serial section, focal collections of ectopic neurons in layer I were found within 0.7 mm of the fused area in all but one sulcus. In all 8 the cerebral cortex beneath the fused sulci showed

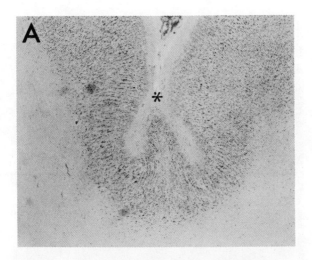

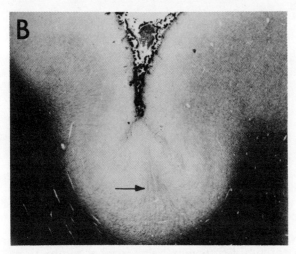

Fig. 6.3 *Typical verrucous dysplasia (*) in case 2:* A, *a Nissl-stained section (note the participation of the upper cortical layers in this malformation);* B, *the adjacent myelin-stained section, with radially directed myelinated fibers in the center of the dysplasia (arrow).*

indistinct lamination and thinning of the cortex, abnormalities similar to those found in the more central parts of the polymicrogyric cortex. The cortical dysplasias, although present in the same areas as the other two types of malformation, were far more numerous in the cingulate and anterior insular cortex, particularly the rostral

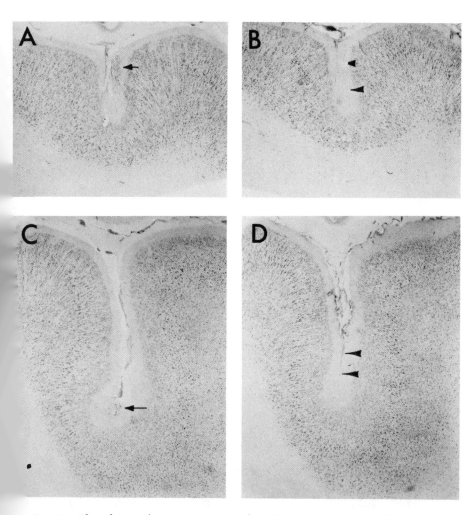

Fig. 6.4 The relation of ectopic neurons in layer I to fused sulci at the edge of polymicrogyric cortex in case 2. A and C, focal accumulations of neurons in layer I (arrows); B and D, fused adjacent gyri within 0.7 mm of the ectopic neurons (arrows).

cingulate cortex (Fig. 6.2B). In the latter two locations focal accumulations of neurons in layer I were not found.

In the thalamus bilateral malformations were present, characterized by disruption of the normal cytoarchitecture and myeloarchitecture of the medial geniculate nuclei and the dorsal parts of the lateral posterior nuclei. The shapes of the medial geniculate nuclei were distorted; large neurons were distributed uniformly throughout, rather than being confined to the usual dorsomedial and anterior positions, and an abnormal band of myelinated fibers was present ventrally. In the nucleus lateralis posterior on each side an aberrant collection of neurons was separated from the main body of the nucleus by an abnormal band of myelinated fibers.

CASE 3 The third brain was that of a right-handed 14-year-old boy who died in his sleep, probably of a cardiac arrhythmia secondary to viral myocarditis. Pregnancy, birth, and early development were considered to have been normal. The boy attended nursery school and kindergarten, where he was described as showing a good adjustment. In the first grade he received speech therapy. In the second grade he was depicted as doing poorly academically, immature, and unable to read. He repeated the second grade with a group of children at a low reading level, but continued to do poorly. In the third grade he was assigned to a reading specialist on a one-to-one basis, and was described as a diligent student with ability to stick to a task. In spite of this his work did not improve, and he was transferred to a parochial school where he attended grades 4 through 6. A diagnosis of severe dyslexia was made, and he received special help with reading. Starting in the seventh grade he attended a special school for the learning-disabled. His death occurred there in the second year.

Extensive testing had been done. The Peabody Picture Vocabulary Test scores in the second and seventh grades were 116 and 122. The WISC scores in the second grade were as follows: verbal IQ of 111, performance of 110, and full-scale IQ of 112. On repeat tests in the third, fifth, and sixth grades he achieved verbal IQs of 98 to 94, performance IQs of 91 to 105, and full-scale IQs of 91 to 100. On the Stanford Achievement Tests in the fourth and seventh grades he showed normal or near-normal performance in vocabulary and listening comprehension. There were marked deficiencies in reading comprehension and spelling (respectively at the 1.3 and 1.0 grade level) in fourth-grade testing. Social science, science, and arithmetic skills were intermediate between these extremes.

Although the patient, like his brother and parents, was predominantly right-handed, he had been ambidextrous for several years. There was no family history of left-handedness. His brother and father were both dyslexic. The brother (who was older) at age 12 years 4 months scored a WISC verbal IQ of 96, performance IQ of 105, and full-scale IQ of 100. The lowest subtest scores were in digit span and coding. Achievement testing at that time showed reading and spelling 2 to 3 years behind expectation.

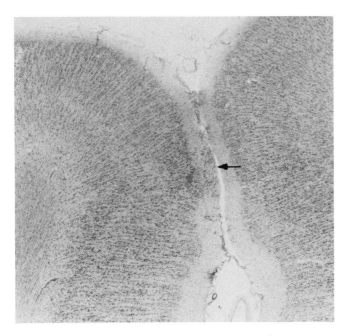

Fig. 6.5 Focal accumulation of neurons in layer I of case 3 (arrow). (Photograph courtesy of Albert Galaburda.)

Analysis of this brain has not yet been completed. The brain weighed 1,597 g after formalin fixation and had no grossly visible abnormalities. The sylvian fissures were of approximately equal length; the planum temporale appeared symmetrical on the two sides. Survey of the serial histological sections showed scattered focal accumulations of ectopic neurons in layer I of the cerebral cortex (Fig. 6.5), confined to the left cerebral hemisphere. Present in greater numbers than in case 2, they were predominantly on the convexity of the hemisphere and involved all lobes. The polymicrogyria observed in case 2 was not present.

Discussion

All three of these brains showed malformations. In cases 2 and 3, studied in whole-brain serial section, the malformations of the cerebral cortex were confined to the left hemisphere. In case 1, described by Drake (1968), bilateral gross lesions were noted at the time the brain was initially cut. It was not stated, however, whether the histological malformations were found in both hemispheres, or if in only one, which it was. Reading difficulties in all three cases were well documented by both school performance and intelligence and achievement tests. All were males with a family history of similarly affected individuals. In our own two cases, the more severely dyslexic subject (case 2) had a more widespread and more severe degree of malformation than case 3, with a milder dyslexia.

In these three brains four different malformations were noted: in the cerebral cortex there were (1) focal accumulations of neurons in layer I (cases 2 and 3), (2) polymicrogyria (case 2 and probably case 1), and (3) focal dysplasias with large neurons scattered from the subcortical white matter into the cerebral cortex (case 2 and probably case 1); furthermore, (4) the thalamus in case 2 showed cytoarchitectonic and myeloarchitectonic abnormalities (case 2). Of these the first two appear to be closely related. The focal accumulations of neurons in layer I correspond to a malformation generally called *verrucous dysplasia*, whose association with polymicrogyria has been noted since its original description by Ranke (1910), who included both lesions in a malformation he called *status verrucosus deformis*. Jacob (1940) called isolated lesions of this type *Hirnwarzen* (brain warts) and Morel and Wildi (1952) used the term *dysgénésie nodulaire disséminée*. Isolated verrucous dysplasia is frequently seen in routine autopsies. Jacob reported them in 11 out of approximately 50 cerebral hemispheres. In these brains they show a striking predilection for the convexity of the frontal lobes (Jacob, 1940; Morel and Wildi, 1952) and a slight predilection for the right cerebral hemisphere. The latter researchers, in a series of 25 brains with these dysplasias, found 56 in right and 48 in left cerebral hemispheres. In their most striking form these malformations appear as grossly visible elevations on the brain surface with a protrusion of the upper cortical layers into the elevation, a central core of radially directed myelinated fibers, and often a large central blood vessel. Smaller lesions show a protrusion of nerve cells from the cortical plate into layer I that often extends to the pial surface. In abnormal brains with polymicrogyria, the verrucous dysplasias often show a close relation to this malformation (Ranke, 1910; Jacob, 1940; Ostertag, 1956).

The configuration of the lesions in cases 2 and 3 conform to these descriptions of verrucous dysplasia and to the illustrations in the papers cited above. In case 2 they were closely associated with the polymicrogyric cortex, appearing at the site of fusion of adjacent gyri. The verrucous dysplasias in these dyslexic brains are unlikely to be a normal finding. In contrast to the dysplasias in allegedly normal brains, these were present in large numbers, were strikingly unilateral, and extended far beyond the convexity of the frontal lobes. It should also be noted that no information had been obtained about developmental disorders of language or behavior in the routine autopsy cases.

The etiology and pathogenesis of both verrucous dysplasia and polymicrogyria has not yet been clearly elucidated from analysis of human autopsy material. There is, however, evidence for the concept that polymicrogyria in humans is the result of a disturbance to the brain during the latter part of the period of neuronal migration (McBride and Kemper, 1982). Experimental support for this view and for a similar pathogenesis for verrucous dysplasia has recently been provided by Dvořák and Feit (1977) and Dvořák et al. (1978) in a combined light-microscopic, Golgi, and autoradiographic study. In the rat, shortly before birth, neurons destined for layers II and III of the cerebral cortex are formed, which complete their migration to the cerebral cortex on postnatal days 1 to 4. Using a freezing lesion at birth, the Dvořák group destroyed the upper cortical layers and showed that a typical microgyric cortex could be produced—the result of the interaction of the area of destruction and of neurons migrating through it to form a superficial cell layer. With more superficial lesions at birth, the same unit produced verrucous dysplasias, polymicrogyric nodules, and shallow, fused microsulci. Lesions at 4 days of age, after the completion of neuronal migration to the cerebral cortex, failed to produce these malformations.

If a similar mechanism were involved in the pathogenesis of verrucous dysplasia and polymicrogyria in dyslexic brains, these anomalies would reflect a disturbance of the left cerebral hemisphere during the later part of the period of neuronal migration to the cerebral cortex. In man this period of migration extends from about fetal week 8 to weeks 16 to 20 (Poliakov, 1937; Sidman and Rakic, 1973; Chatel, 1976).

Further support for this timing is provided by the pattern of focal dysplasias found in case 2 and possibly case 1. These lesions, which correspond to similar lesions described by Taylor et al. (1971), have as one of their characteristics ectopic neurons scattered in the subcortical white matter. These neurons indicate that the time of formation of the dysplasias must have been before neuronal migration to the cerebral cortex was finished. Their usual areas of predilection, the cingulate cortex and anterior insula, are the last areas of the cerebral cortex to show completion of neuronal migration (Poliakov, 1937; Chatel, 1973).

The thalamic malformations in case 2 were bilateral and confined to two thalamic nuclei, the medial geniculate and the nucleus lateralis posterior. This combination of bilateral thalamic lesions and unilateral cortical malformations is intriguing and, like polymicrogyria and verrucous dysplasia, difficult to understand without an appropriate animal model. Galaburda and Eidelberg (1982), in their analysis of the thalamic lesion in this case, propose an interesting hypothesis. It is based on the observation by Rakic and Sidman (1969) in humans of a possible common origin, from the same germinal zone (the ganglionic eminence), of neurons destined for both the pulvinar-lateralis posterior thalamic complex and connectionally related homotypical neocortical association areas. Galaburda and Eidelberg postulate that the cells for both the lateralis posterior nucleus and its cortical projection area, the inferior parietal lobule, may have originally come from a common pool in this germinal zone and thus shared a defect in migration to their target areas. To account for the bilaterality of the two thalamic lesions, they postulated that neurons migrate to both thalami from the germinal zone. A similar mechanism was suggested for the medial geniculate nucleus and its related auditory cortices, which in the left cerebral hemisphere were involved in the polymicrogyric malformation and verrucous dysplasias.

Alternatively, and in harmony with the postulated etiology of the lesions in the neocortex, the possibility exists that these thalamic malformations may reflect an insult during the same period in which the cortical malformations have their origin. According to Yakovlev (1969), the germinal zone for the thalamic neurons lining the third ventricle is exhausted by 13 weeks. From 6 weeks until 13 weeks the thalamus contains many "undifferentiated neuroblasts." The thalamocortical projections appear from the 8th until about the 15th week of gestation. The last to appear are the posterior thalamic radiations from the dorsolateral zone (which includes the nucleus lateralis posterior) and the posterior zone of the thalamus, and from the geniculate bodies. An injury to the medial geniculate nuclei and the nucleus lateralis posterior at this late stage, a time corresponding to the end of neuronal migration to the cerebral cortex, might account for both the abnormal cytoarchitecture and the aberrant myelinated fiber bundles.

Included in this analysis are only the readily observable lesions in these brains, that is, the malformations. More subtle defects are likely to be present. On the other hand, malformations may exist in some cases which affect the regional vascular architecture as well as the cytoarchitecture. We have had occasion to observe a middle-aged, left-handed, dyslexic lawyer who presented with partial complex seizures and was found to have an arteriovenous malformation involving the left temporoparieto occipital region. Two additional dyslexic males whose brains have been made available, died of spontaneous subarachnoid hemorrhages; their brains will undergo detailed analysis. Levine et al. (1981) described a case of acquired dyslexia in a child who at a young age had a temporal lobectomy for hemorrhage, which we believe arose in a vascular anomaly present in a malformed left temporal cortex. Only with continued observation of additional cases, and further refinement of such observations, will we be able to determine which abnormalities are responsible for the clinical manifestation of dyslexia. Regardless of whether only one of these, or a combination, is causally related, it is evident that malformation is a common feature of the brain morphology of dyslexia. Three out of three postmortem cases have shown such lesions. The available evidence suggests that they are the result of an abnormal developmental process toward the end of the first half of pregnancy. The findings in the cerebral cortex of the two most thoroughly studied cases suggest that the left hemisphere is especially vulnerable.

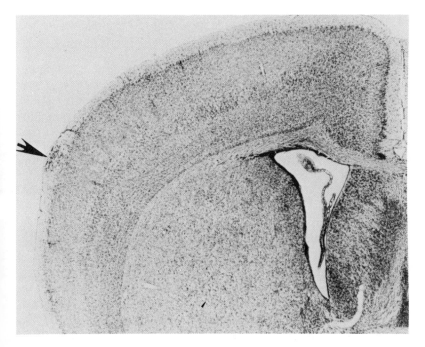

Fig. 6.6 *Focal accumulation of neurons in layer I (arrow) of the New Zealand Black mouse, a model of autoimmune disease. (Photograph courtesy of Gordon Sherman and Albert Galaburda.)*

Of interest in this regard is the role postulated by Geschwind and Behan (1982) of sex hormone effects on left hemisphere development, and their report of clustering in related individuals of left-handedness, dyslexia and other developmental disabilities, autoimmune disease, and migraine. Experimental support for this relation to autoimmunity is provided by a recent observation of Sherman and Galaburda (personal communication) of typical verrucous dysplasias in a mouse model of autoimmune disease, the New Zealand Black mouse (Fig. 6.6). In these animals the lesions thus far observed occur in either cerebral hemisphere, but usually not both.

Since the cases of dyslexia cited in this chapter are all familial, genetic factors are likely to play some role, either in direct control of the abnormal developmental processes establishing susceptibility of the left hemisphere to altered hormonal balance or in conferring a tendency to high production of hormones.

References

Chatel, M. 1976. Développement de l'isocortex du cerveau humain pendant les périodes embryonnaires et fétales jusqu'à la 24ème semaine de gestation. *J. Hirnforsch.* 17:189–212.
Critchley, M. 1964. *Developmental Dyslexia*. London: Heinemann Medical Books.
Drake, W. E. 1968. Clinical and pathological findings in a child with a developmental learning disability. *J. Learn. Dis.* 1:486–502.
Duffy, F. H., Denckla, M. B., Bartels, P. H., and Sandini, G. 1980. Dyslexia: regional differences in brain electrical activity by topographic mapping. *Ann. Neurol.* 7:412–420.
Dvořák, K., and Feit, J. 1977. Migration of neuroblasts through partial necrosis of the cerebral cortex in newborn rats. *Acta Neuropathol.* 38:203–212.
Dvořák, K., Feit, J. and Juránková, Z. 1978. Experimentally induced focal micropolygyria and status verrucosus deformans in rats: pathogenesis and interrelations. *Acta Neuropathol.* 44:121–129.
Galaburda, A. M., and Eidelberg, D. 1982. Symmetry and asymmetry in the human posterior thalamus. II. Thalamic lesions in a case of developmental dyslexia. *Arch. Neurol.* 39:333–336.
Galaburda, A. M., and Kemper, T. L. 1979. Cytoarchitectonic abnormalities in developmental dyslexia: a case study. *Ann. Neurol.* 6:94–100.
Geschwind, N., and Behan, P. 1982. Left-handedness: association with immune disease, migraine, and developmental learning disorder. *Proc. Natl. Acad. Sci. USA* 79:5097–5100.
Haslam, R. H., Dalby, J. T., Johns, R. D., and Rademaker, A. W. 1981. Cerebral asymmetry in developmental dyslexia. *Arch. Neurol.* 38:679–682.

Hier, D. B., LeMay, M., Rosenberg, P. B., and Perlo, V. P. 1978. Developmental dyslexia: evidence for a subgroup with reverse asymmetry. *Arch. Neurol.* 35:90–92.

Hinshelwood, J. 1895. Word-blindness and visual memory. *Lancet* 2:1564–70.

Hinshelwood, J. 1917. *Congenital Word Blindness.* London: Lewis.

Jacob, H. 1940. Die feinere Oberflächengestaltung der Hirnwindungen, die Hirnwarzenbildung und die Mikropolygyrie. *Z. Neurol. Psychiat.* 178:64–84.

Levine, D. N., Hier, D. B., and Calvanio, R. 1981. Acquired learning disability for reading after left temporal damage in childhood. *Neurology* 31:257–264.

McBride, M. C., and Kemper, T. L. 1982. Pathogenesis of four-layered microgyric cortex in man. *Acta Neuropathol.* 57:93–98.

Morel, F., and Wildi, E. 1952. Dysgénésie nodulaire disséminée de l'écorce frontale. *Rev. Neurol.* 87:251–270.

Morgan, W. P. 1896. A case of congenital word-blindness. *Brit. Med. J.* 2:1378.

Orton, S. T. 1925. "Word-blindness" in school children. *Arch. Neurol. Psychiat.* 14:581–615.

Ostertag, B. 1956. Die Einzelformen der Verbildungen (einschliesslich Syringomyelie). In O. Lubarsch, F. Henke, and R. Rossle, eds., *Handbuch der speziellen pathologischen Anatomie und Histologie*, vol. 13, pp. 363–601.

Poliakov, G. I. 1937. *Early and Intermediate Ontogenesis of the Human Cerebral Cortex.* (In Russian.) Moscow: Brain Institute–Academy of Medical Sciences.

Rakic, P., and Sidman, R. L. 1969. Telencephalic origin of pulvinar neurons in the human fetal brain. *Z. Anat. Entwickl. Gesch.* 129:53–82.

Ranke, O. 1910. Beiträge zur Kenntnis der normalen und pathologischen Hirnrindenbildung. *Beitr. Path. Anat.* 47:51–125.

Rosenberger, P. B., and Hier, D. B. 1980. Cerebral asymmetry and verbal intellectual deficits. *Ann. Neurol.* 8:300–304.

Sidman, R. L., and Rakic, P. 1973. Neuronal migration with special reference to developing human brain: a review. *Brain Res.* 62:1–35.

Taylor, D. C., Falconer, M. A., Burton, C. J., et al. 1971. Focal dysplasias of the cerebral cortex in epilepsy. *J. Neurol. Neurosurg. Psychiat.* 34:369–387.

Yakovlev, P. I. 1969. Development of the nuclei of the dorsal thalamus and of the cerebral cortex. In S. Locke, ed., *Modern Neurology.* Boston: Little, Brown, pp. 15–53.

Part Two

Brain Asymmetry in Other Species

Chapter 7

Learning, Forgetting, and Brain Repair

Fernando Nottebohm

During learning there are changes in the performance of neurons and of the networks they form (Fifkova and Van Harreveld, 1977; Alkon and Crow, 1980; Kandel and Schwartz, 1982). These altered states constitute memories. The number of networks that can be altered in this manner presumably sets a limit to how much can be learned and remembered. In this sense, learning requires network space. New network space can be made available in adulthood, by replacing used network components with new ones. Maintenance of learning potential, therefore, may require forgetting. The relation between learning and forgetting may provide useful insight into the natural processes of brain repair. Simply put, the study of brain processes for learning is also the study of brain processes for self-repair.

To develop this theme, I shall focus on the song-learning system of canaries and other songbirds. The history of this system is unusual. Its roots are in ethology (Thorpe, 1958, 1961; Marler and Tamura, 1964; Marler, 1970a; Marler and Peters, 1977; review in Nottebohm, 1980a, 1984). To date it remains the only vertebrate system for which a relation has been established between a naturally occurring learned behavior and the brain pathways that control it.

Song learning is the acquisition of a song repertoire by reference to auditory models. We are all familiar with this technique. When we learn to speak, or learn new languages, we do so by imitating the sounds we hear. Birds do not speak, but sing. They sing their species and sex, their individual identity, marital status, stamina, and territorial ownership (Marler, 1956). These are important messages. The response they elicit from conspecific males and females determines the sender's ability to contribute to future generations. For most orders of birds the song that proclaims this information is defined by the genome, unaltered by individual experience. But in three cases song is learned, so that one individual can mimic the sound of another or can improvise on a diversity of sounds within a general, species-typical plan. This has happened in songbirds (Passeriformes), parrots (Psittaciformes), and hummingbirds (Apodiformes); we do not know the evolutionary history of these special cases (Nottebohm, 1972, 1975).

Acquisition of Song

STAGES IN SONG LEARNING   Vocal learning requires that a sound be perceived and remembered, or that a particular sound be expected and recognized when heard. During song learning, vocal output is modified until it matches the remembered or expected model. One might guess that this represents a process or trial and error for each sound learned, but that does not seem to be the case. Birds that learn their song and calls go through a "subsong" stage, which Charles Darwin likened to the babbling of infants (Thorpe, 1958). The sounds produced do not require a social context, are highly variable, and delivered at very low volume, often as the young bird seems to doze. The impression one gets is that during subsong the young bird learns how to play its musical instrument; later it learns what to play. Learning to play the vocal organ, or syrinx, may mean that particular sounds are mapped onto particular motor acts. Once this code is available, any model that falls within the range of sounds produced by the syrinx can be imitated. This is not, however, the only constraint on vocal learning. Many species bring to the learning task a predisposition to recognize, and imitate, conspecific song; songs of other species are ignored (Thorpe, 1958; Marler and Tamura, 1964; Marler and Peters, 1977). Unlearned motor rules may also discourage excessive drift in the production of species-typical learned song (Price, 1979). For example, canaries deafened before the onset of singing produce song that is simpler than that of intact birds, yet with a similar pattern of delivery (Marler and Waser, 1977; Güttinger, 1981). Even the song of mockingbirds, rich in components borrowed from other species (Borror and Reese, 1956; Howard, 1974; Baylis, 1983), conforms to a well-defined, mockingbird-typical pattern of delivery. What we have, then, is a learned behavior guided and constrained during its ontogeny by unlearned motor and perceptual rules.

TEMPORAL RESTRICTIONS ON LEARNING In many species song learning occurs early in ontogeny, during the first weeks or months after hatching. Appropriate models heard during this critical period are imitated. Once the critical period ends, new models cease to have an effect on vocal output (Thorpe, 1958; Marler and Tamura, 1964; Immelmann, 1969). The existence of critical periods is somewhat paradoxical, since it means that even as the individual masters its musical instrument and learns its first score or scores, it also loses the ability to add to the repertoire.

In the zebra finch, the critical period ends by 90 days after hatching. Prior to that time male offspring learn to imitate the song of the father or foster father; these birds live for several more years, but their song remains virtually unchanged (Immelmann, 1969; Price, 1979). In white-crowned sparrows, the song model is learned during the first 2 months, although singing does not start until later (Marler and Tamura, 1964; Marler 1970b).

Among white-crowned sparrows (genus *Zonotrichia*), it has been argued, critical periods for song learning help promote close breeding between members of local populations. For this to happen, both males and females must learn at an early period the songs common to their natal grounds. While males use this information to develop their own song, females use it to choose their partners (Nottebohm, 1969a; Baker et al., 1981). This example is of particular interest, since it shows that even though females do not normally sing, they can learn and remember song models.

In contrast to these examples, cardueline finches are open-ended learners that can modify their song and call repertoire in successive years (Mundinger, 1970). The canary is a member of this group. It too can continue to learn new song repertoires (Nottebohm and Nottebohm, 1978). Understanding the brain correlates that make for either critical-period or open-ended learners would add much to our insight into the kinds of variables that favor or hinder learning of new sensory-motor integrations. Our only clues so far are that gonadal hormones probably play a role in determining the end of the critical period (Nottebohm, 1969b; Kroodsma and Pickert, 1980).

Neural Mechanisms of Song

LEFT HYPOGLOSSAL DOMINANCE Song learning in canaries is a left-sided affair. The vocal organ, or syrinx, consists of two fairly symmetrical halves. Each syringeal half has its own air supply, muscular control, and innervation, and can, on its own, produce sound. The tracheosyringeal branch of the left hypoglossus nerve innervates the muscles of the left syringeal half, while the muscles of the right side are innervated by the tracheosyringeal branch of the right hypoglossus. Section of the right hypoglossus in adult canaries leads to the loss of few song syllables, if any. Section of the left hypoglossus is followed by the elimination of 90% or more of all song syllables (Nottebohm and Nottebohm, 1976). From this it has been inferred that most sounds of canary song are produced under left hypoglossal control. Since song learning requires that output be modified by reference to auditory feedback (Konishi, 1965), we may assume that the left hypoglossal motor neurons, and the higher centers that control them, learn to sing. What, then, is the role of the right hypoglossus — beyond the few syllables it contributes to song?

It has been reported that the gross patterns of activity in the left and right tracheosyringeal nerves do not differ during song production (McCasland, 1983). This observation need not be in conflict with the observed left hypoglossal dominance. Sound is produced by periodic changes in air pressure: these can occur in either syringeal half, even after denervation, as air flow is modified by the Bernoulli effect and turbulence (Nottebohm et al., 1979). Muscular control of the dynamics of air flow through either syringeal half can, then, produce either song or silence. At the tracheosyringeal level, commands to produce song (left side) or silence (right side) might look similar as they overlap during the same expiratory pulse.

The song of adult intact canaries normally includes some syllables that are accompanied by simultaneous, poorly modulated pulses of sound, or noise. As canaries go from year 1 to year 2, the size of the syllable repertoire increases, while the percentage of syllables accompanied by noise decreases (Nottebohm et al., 1979). Thus, learning to sing is also learning to silence.

Despite the observed functional asymmetry the right syringeal half can also produce canary song if the left hypoglossal nerve is cut before the onset of song learning. If the same operation is done in adulthood, the right syrinx develops a new, albeit smaller, song repertoire. We may infer that while respiratory needs are well served by a symmetrical syrinx, half a syrinx suffices for singing.

OVERLAP BETWEEN NEURAL ENCODING AND DECODING   The greatest advantage offered by the song-control system of songbirds is that a series of discrete nuclei control a discrete, quantifiable, natural, learned behavior. This may result from the fact that sound production resides in a specialized organ, the syrinx, whose main function is to sing.

The highest forebrain song-control nucleus is the hyperstriatum ventrale, pars caudalis (HVc). As shown in Fig. 7.1, HVc projects to a second forebrain nucleus, the robustus archistriatalis (RA). RA projects to the hypoglossal motor neurons that innervate the trachea and syrinx (Nottebohm et al., 1976). HVc is particularly interesting in that it receives auditory input, as might be expected from a nucleus involved in song learning (Kelley and Nottebohm, 1979). HVc neurons respond to a diversity of sounds, including song (Katz and Gurney, 1981; McCasland and Konishi, 1981), and can do so in a very selective manner (Margoliash, 1983). They also fire during song production (McCasland and Konishi, 1981). Thus, HVc may have a distinct role in the encoding and decoding of song.

The motor theory of speech perception predicted that phonetic decoding would be possible only via a system that knew how phonemes were produced, so that a sound heard could be related to how that sound was produced (Liberman et al., 1967). In consonance with this view, electrical stimulation of Broca's area for speech control in the human frontal lobe has been shown to alter aspects of both speech encoding and decoding (Ojemann and Mateer, 1979). There is, then, a tempting functional parallel between HVc and Broca's area. HVc projects to RA, which has been likened functionally to layer 5 of motor cortex in mammals (Nottebohm et al., 1976). Afferents to HVc and RA have been described, so that at least anatomically, a good deal is known about the circuitry involved in song control (Nottebohm et al., 1982).

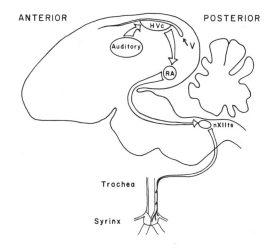

Fig. 7.1 Schematic sagittal section through the brain of an adult male canary, showing ipsilateral connections from the hyperstriatum ventrale, pars caudalis (HVc), to the nucleus robustus archistriatalis (RA), and from RA to the hypoglossal motor neurons that innervate tracheal and syringeal muscles (nXIIts). The auditory input to nucleus HVc is also shown. The space above HVc is the forebrain ventricle (V).

HEMISPHERIC DOMINANCE AND ITS REVERSAL The left HVc of canaries projects to the left RA, which in turn projects to the left hypoglossus, which innervates the left syringeal half. The same is true for the right side (Nottebohm et al., 1976). This predominance of ipsilateral connections has been confirmed physiologically (Arnold, 1980; Paton and Manogue, 1982). Not surprisingly, then, lesions to the left HVc are far more damaging to song control than lesions to the right HVc. Following left HVc destruction, the quality of song control approximates that in subsong. It is as if a whole learned program were taken away. Destruction of the right HVc leads to loss of some syllables and modification of others, but much song remains as before the operation (Nottebohm et al., 1976; Nottebohm, 1977).

Hemispheric dominance for song control can be reversed in adult canaries. In the months after destruction of the left HVc a new repertoire is developed, this time under right HVc and syringeal control. In this sense, the two sides seem to be equipotential (Nottebohm, 1977). If an adult canary is deafened immediately before left HVc destruction, it develops a deaf-type, simple song. This song is not affected by subsequent right hypoglossal section (Nottebohm et al., unpublished observations). It seems as if the right HVc has to hear a deficit before it is able, along with the right syringeal half, to assume a dominant role in song production.

It seems possible that left and right song-control pathways complement each other in the production and perception of song. The need for this complementarity is suggested by the following observation. Neurons that fire during song production remain refractory to sound input for several seconds after a sound has been produced (McCasland and Konishi, 1981). Parallel pathways for perception and production would allow simultaneous occurrence of both functions, which would seem important for song learning. It is not known whether canaries with a single HVc (right or left) can imitate external models. Although right and left song-control pathways seem to have different roles, anatomically they are equally well developed (Nottebohm et al., 1981). No direct anatomic pathway linking right to left HVc has yet been described.

BRAIN SPACE FOR A LEARNED SKILL There is a threefold range in the size of nucleus HVc and RA in our close-bred Waserschlager canary population. There is also a threefold range in the size of adult song repertoires, where size stands for the number of syllable types produced. These two variables are positively related in a significant manner. Male canaries with large song repertoires tend to have large

HVc's and large RA's; male canaries with small HVc's and small RA's tend to have small song repertoires (Nottebohm et al., 1981). A similar relationship has been observed in zebra finches (Nottebohm and Crane, unpublished observations) and in marsh wrens. The latter example is of particular interest. The same species of marsh wren (*Cistothorus palustris*) occurs throughout the United States. There is, however, considerable variability in song complexity among populations. California populations have song repertoires that are three times as large as those recorded in New York's Hudson Valley. Although the bodies and brains of the western birds are slightly smaller than those of their eastern counterparts, the HVc and RA are 40% and 30% larger, respectively, in the western birds (Canady et al., ms). We may conclude, then, that the extent of network space for a learned skill and the amount of that skill that is learned are related. The causality and direction of this relationship have not yet been established.

CREATION OF EXTRA NETWORK SPACE The account thus far points to a system with functional asymmetry and adult plasticity, in which network space may play a role. How is plasticity controlled? Several lines of evidence point to gonadal hormones. First of all, nuclei HVc and RA are larger in males than in females. Neurons in the larger male RA have dendrites that are longer than dendrites in the corresponding female cell type (DeVoogd and Nottebohm, 1981a). Part of this difference may result from hormonal influences early in ontogeny, as has been shown in the zebra finch (Gurney and Konishi, 1980; Gurney, 1981). But there is a role, too, for adult hormonal effects.

Female canaries do not normally sing, but females treated with physiological doses of testosterone develop male-like song, also a marked increase in the size of HVc and RA. The anatomical changes involved in this increase have been documented (Nottebohm, 1980b; DeVoogd et al., 1982).

Systemic testosterone treatment of adult female canaries induces song 2 to 5 days later. Two to 3 weeks after onset of hormone treatment the song of those females has reached maximal complexity and become stable (unpublished observations). The RA of testosterone-treated females sacrificed 4 to 5 weeks after onset of treatment is 70% to 90% larger than that of control birds which have been treated with cholesterol-filled or empty Silastic spheres (Nottebohm, 1980b; DeVoogd et al., 1982). This increase in volume does not result from the addition of new neurons (Goldman and Nottebohm, 1983) but reflects, in part at least, dendritic growth.

The dendrites of a particularly well defined, sexually dimorphic neuronal type are 49% longer after testosterone treatment than in controls (DeVoogd and Nottebohm, 1981b). This increased length is accompanied by a net gain of 70% in the number of RA synapses (DeVoogd, et al., 1982). Since nucleus RA is the obligatory exit for forebrain motor pathways controlling song, the testosterone-induced synaptogenesis probably represents changes in circuitry relevant to the newly acquired behavior. Research is under way to identify the sources of these new inputs.

GROWTH OF NETWORKS AS LEARNING PROCEEDS The effect of testosterone on the HVc and RA of females is somewhat puzzling, since females normally do not sing. Does this consequence relate to natural events? Probably so, in males. To make this clear, it helps to review the ontogeny of their song-control system.

The brains of young male canaries reach adult size by day 15 after hatching. At that time forebrain song-control systems are very poorly developed. HVc and RA reach full adult size 7 to 8 months after hatching. A month after hatching, HVc is one-eighth adult size and RA one-third. The first stage of song development, subsong, starts some 40 days after hatching and lasts for about 2 weeks. The size of HVc virtually quadruples and that of RA doubles from day 30 to day 60 after hatching. The rate of growth slows thereafter, as the young bird continues to add new syllable types to its song repertoire. By early spring the first year's song repertoire has become stable and remains so for the duration of the breeding season.

SEASONAL FLUCTUATION OF SONG-CONTROL NUCLEI As the breeding season ends, song becomes less frequent and more variable. This quality of variability is reminiscent of the "plastic song" stage in development. During mid or late summer most adult male canaries become totally silent and remain so for about a month. Late in the summer song recommences, again in a variable manner, and it is at this time that old syllables are modified and new ones added, with a net gain in the total number of syllables. This new repertoire is again stable by the onset of the following breeding season. HVc and RA sizes show no significant change from the spring of year 1 to the spring of year 2. In late summer, however, after the end of the first breeding season, HVc and RA are approximately half as large as they were in the preceding spring. In late summer testis volume is 1/140 of what it had been in the spring, and blood androgen levels are close to zero. All these decrements have been made up by the following spring. It is tempting to conclude that the hormone-induced changes observed in females happen also in males, where they occur in a reversible manner and result in the acquisition of a new song repertoire (Nottebohm, 1981).

Neurogenesis in Adulthood

The extent of seasonal and hormone-induced changes in nucleus HVc caused us to consider the possibility that network space was added not just in the form of new dendrites and new synapses, but that new neurons might be added too. At the time, we were aware that adult neurogenesis had been demonstrated in the olfactory mucosa of mammals (Graziadei and Monti-Graziadei, 1979). We did not realize that it had also been shown in the olfactory bulb, dentate gyrus, and visual cortex of normal adult rodents (Kaplan and Hinds, 1977; Kaplan, 1981; Bayer et al., 1982).

It was our expectation that if neurogenesis was involved in the changes of HVc size, it would be governed by gonadal hormones, possibly testosterone. We used year-old females with prior reproductive experience, divided into two groups. Birds in one group were treated with testosterone; those in the other were treated with cholesterol. Radioactively labeled thymidine, a marker of DNA synthesis, was administered systemically during subsequent days. Birds in both groups were sacrificed one month after they had received the tritiated thymidine.

Two surprises were in store for us. First, as many as 1.5% of all HVc neurons were labeled per day of [^{3}H] thymidine treatment. Second, there was no difference in the labeling indexes of the testosterone-treated and the cholesterol-treated birds. A similar labeling index was observed when [^{3}H] thymidine was given 14 days prior to onset of the testosterone treatment. A subsequent experiment showed that the new neurons were produced in the ventricular zone lining the telencephalic ventricle dorsal to HVc. Interestingly, there were no labeled neurons in RA (Goldman and Nottebohm, 1983).

Since the size of nucleus HVc does not change in adult females from the spring of year 1 to the spring of year 2, we concluded that the recruitment of new neurons must be accompanied by neuronal death. Otherwise the number of HVc neurons would double over a 50-day period. We inferred, then, that we were witnessing a replacement of "old" neurons by "new" neurons.

THE FUNCTION OF NEW NEURONS What seemed particularly intriguing to us was that this phenomenon should occur in the vocal control nucleus of adult female canaries. Females, as indicated earlier, normally do not sing, but they probably remember the songs of males they hear, as has been shown for females of another songbird species (Baker et al., 1981). Since we know that HVc has access to auditory information, we wondered whether the main role of this nucleus in females could be song recognition. If so, the production and replacement of neurons in HVc would be related to perceptual, rather than to motor, learning. This, of course, was quite different from our original hypothesis.

The perceptual or motor role of new HVc neurons in female canaries could not be satisfactorily established because, as has been shown for other carduelines, adult females may continue to alter their call repertoires (Mundinger, 1970), a phenomenon that is in principle not different from song learning. The obvious way to test the hypothesis that neurogenesis need not be related to motor learning was to look for it in a "critical period" species such as the zebra finch, after stereotyped adult song had been acquired. If neuronal replacement were related to song recognition—a perceptual phenomenon—then it would continue to occur after the end of the critical period.

Two adult male zebra finches were treated with [³H] thymidine well after the end of the critical period. In these birds new neurons were recruited into nucleus HVc at a rate of 0.26% per day (Nottebohm and Kasparian, 1983). This rate is comparable to that observed in male canaries treated with [³H] thymidine at the same time. At this rate the number of neurons in the HVc would double in about 300 days. It seems fair to conclude that neurogenesis in a song-control nucleus may be related to the updating of either perceptual or motor memories.

THE REBUILDING OF A FOREBRAIN Our initial questions focused on nucleus HVc and song control, but the evidence and issues involved transcend this system. The occurrence of labeled neurons following [³H] thymidine treatment in adulthood is not restricted to nucleus HVc. Rather, such cells are found (though at lower densities) in various parts of the forebrain, but this pattern of distribution does not occur in animals sacrificed a short time later. In this case most of the label, other than that found in glia or endothelial cells, occurs in cells of the ventricular zone abutting the forebrain ventricles. Had this labeling been simply a marker of DNA repair, it would have occurred in situ rather than in the ventricular zone. As in the case of HVc, new neurons in the rest of the forebrain seem to arise from ventricular zone stem cells, and to migrate from there. In the canaries processed so far we have found only three instances of labeled neurons in the midbrain; conceivably they could have reached there after starting from forebrain sources. Although our analysis is far from complete, it is striking that whereas many labeled neurons occur in the forebrain, virtually none have been found in the septum, diencephalon, mesencephalon, cerebellum, or medulla. Our tentative hypothesis is that in birds the phylogenetically old parts of the brain make little use of adult neurogenesis. It is in the evolutionarily more modern forebrain, to which control of the more flexible behaviors and learning are ascribed, that neurogenesis seems best represented.

Forebrain neurogenesis in adulthood has now been observed in male and female canaries, male and female zebra finches, male and female parakeets (Manogue and Nottebohm, unpublished observations), and male and female doves (Cohen and Nottebohm, unpublished observations). These birds represent three separate avian orders — Passeriformes, Psittaciformes, and Columbiformes. It seems fair to assume that the phenomenon is widely distributed among birds.

A caveat is in order here. The neuronal identity of new HVc neurons was established by use of autoradiography on semithin sections, 1 μm thick. Adjacent sections were prepared for transmission electron microscopy. Some of the cells identified as labeled neurons under the light microscope were described ultrastructurally and confirmed to be neurons (Goldman and Nottebohm, 1983). Putative neurons elsewhere in the forebrain have not yet been studied in this way, but their large sizes, round clear nuclei, and central nucleoli — traits shared with the HVc neurons — make it extremely likely that our identification is correct. If this cell type were some form of glia, we would expect to see it labeled throughout the brain.

Solutions to a Shortage of Learning Space

If memory space is limited, yet the need for new memories is great, then strategies may have evolved to meet this requirement. We have focused on a cluster of four brain traits — laterality, network space, synaptogenesis, neurogenesis — that seem related to learning. Let us have another look at these traits.

LATERALITY Lateralization of function has been much touted as a strategy for doubling the size of memory banks and hence the brain's potential for processing information. The classical example is hemispheric dominance for speech and language skills in humans

(Penfield and Roberts, 1959; Sperry, 1974). The usual concept is, of course, that right and left homologous parts of the forebrain can diverge during development so as to serve very different functions. This notion is attractive, to some extent probably true, but is not without its problems.

Network function is determined, presumably, by the nature of inputs and outputs. Does this mean that when hemispheres differ in function they differ in connectivity? Yes and no. Right or left hemidecortication before the onset of speech does not preclude the development of normal speech, manifested by normal phonetic encoding and decoding (Dennis and Whitaker, 1976). To this extent, each hemisphere may have the full potential to develop speech. If so, what is the fate of the speech-control network in the subordinate hemisphere of intact brains? It has been argued that it controls prosody, the intonation and affective color of speech sounds (Ross and Mesulam, 1979). There is, however, no evidence that prosodic and phonetic functions cannot coexist in the same hemisphere. There has been no mention of abnormal prosody in patients subjected to hemidecortication before the onset of speech development (Dennis and Whitaker, 1976).

Sperry (1982) admitted to being puzzled by the fact that speech comprehension could be so upset by unilateral lesions to language areas of the dominant hemisphere, when comprehension could be shown to be present to some extent in both hemispheres after surgical section of the corpus callosum in adulthood for the treatment of long-standing epilepsy. The possibility thus arises that for some functions hemispheric dominance results not so much from the different skills assumed by homologous pathways as from suppression of a subordinate pathway, which may then remain underused. Underuse may stem from incompatibility of attentional systems within a given hemisphere (Kinsbourne and Cook, 1971; Okazaki-Smith et al., 1977) or from the fact that a democratic two-headed system makes for trouble, as has been often suggested as an explanation for stammering (Orton, 1937).

Some of the functional asymmetries of the human brain may depend on anatomical asymmetries (Geschwind and Levitsky, 1968; LeMay, 1976; Galaburda et al., 1978). In cases of marked asymmetry, as reported for the planum temporale (Geschwind and Levitsky, 1968), it seems reasonable to assume that homologous parts of the two hemispheres have become so different as to be incapable of being functionally alike. Such anatomical differences may, for example, underlie hemispheric differences in syntactic skill (Dennis and Kohn, 1975; Dennis and Whitaker, 1976). Still, it may be too early to decide how much of hemispheric dominance is due to anatomical asymmetries and how much to suppression and underuse. If homologous pathways remain underused in the subordinate side of the adult brain, this should be of great interest to neurologists as they consider prognosis and treatment. The avian material presented here does not yet permit a choice of hypothesis between "underuse" and "complementary specialization." Our fragmentary knowledge of the song-control system suggests the presence of anatomic symmetry and functional asymmetry. In time, as the interaction and specific functions of right and left pathways for song control are better understood, they may well encourage a reevaluation of lateralization of function in the human brain.

NETWORK SPACE The issue of laterality is an issue of network space. Yet it may not be reducible to mere space. I rather suspect that the basic bilateral symmetry of the brain may have some built-in redundancies and inevitable underuse. This is quite unlike the economy of network space found when comparing sexes. The sex that sings has song-control nuclei several times larger than those of the sex that does not sing (Nottebohm and Arnold, 1976). The same principle of economy, relating size of brain nuclei to learning ability, seems to apply to variability within populations and between populations, and it can also be discerned during ontogeny. From a practical viewpoint these observations are encouraging. HVc and RA space can be reduced to number and type of neurons and synapses. This brings me to synaptogenesis, and its relation to learning.

SYNAPTOGENESIS The contribution of synaptogenesis to the normal functions of adult brain circuits has received much recent attention (Cotman and Nieto-Sampedro, 1982; Nieto-Sampedro et al., 1982). It has been proposed that the adult complement of synapses arises during ontogeny by a process of functional selection, whereby some connections become stable and others degenerate (Changeux et al., 1973; Changeux, 1974).

Views of the fate of synapses present in adulthood vary. Eccles (1973) thought that functional modulation of existing synapses probably sufficed to provide the adult brain with all the flexibility needed for learning. This theme has been taken up in research on learning in *Aplysia*, in which existing synapses are seen as plastic brokers of information (Kandel, 1978; Kandel and Schwartz 1982). In the view favored by the Kandel group, this plasticity comes about by the modulation of presynaptic calcium channels; amounts of intracellular ionic calcium are thought to regulate transmitter release and so govern the efficacy of synaptic transmission. This mechanism has been proposed as an explanation of habituation, sensitization, and associative learning. A different hypothesis offers a potentially complementary mechanism (based on studies in the hippocampus). That is, postsynaptic plasticity is suggested as the variable that regulates learning. In this theory the numbers of transmitter receptors (Baudry et al., 1980) or of spine configurations (Fifkova and Van Harreveld, 1977) are thought to change as a result of increased use and thus determine future synaptic efficacy.

Are synapses permanent, once formed and confirmed by use? Changeux (1974) suggests that there are learning periods when the motility of some nerve terminals gives them the ability to establish transiently a multiplicity of contacts. Subsequent electrical activity stabilizes some synapses and eliminates others. Changeux sees such a process of selection as being responsible, for example, for the learning of a particular language from the many that are possible. He is attempting to account for how much specificity must be provided by the genome and how much shaping can be provided by experience.

Cotman and Nieto-Sampedro (1982) take a somewhat different view. They suggest, first, that synaptic growth — and thereby effectiveness — can be induced by changes in neuronal activity, which may track natural stimuli. Second, they suggest that in some parts of the brain (the hippocampus, for instance) synapses are constantly formed and unformed. Part of this ongoing process may reflect changing patterns of use. These authors however, seem to believe that in some parts of the brain synaptic turnover occurs as part of an inexorable cycle of birth, growth, and breakup (Nieto-Sampedro et al., 1982). Carlin and Siekevitz (1983), reviewing the evidence on synapse plasticity, conclude that in many parts of the brain, in periods when learning occurs, a subset of existing synapses undergoes division. Where previously contacts between two neurons were represented by 1,000 synapses, for example, they are now represented by 2,000. In this manner the influence of one neuron on another would be strengthened considerably, and the information it conveys would gain in salience.

Synaptogenesis is interesting in another context. Suppose that presynaptic and/or postsynaptic changes brought about by use are irreversible. This would mean that as synapses partake in learning, an increasing fraction would become permanently biased. If the same neurons are to be used for learning new information, "used" synapses must be replaced by "fresh" ones. During evolution brains may have been able to choose between the coding of memories by means of either reversible or irreversible synaptic changes. Changes that are temporary, once forgotten, make the pathway available for new learning; changes that are permanent preserve memories, but at the expense of an ever-dwindling population of plastic synapses. In the latter case a rejuvenation of pathway potential could still be achieved by synapse replacement, with consequent loss of memory. New synapses, in turn, may recreate the connectivity of the synapses they replace, or they may be novel contacts that alter the balance of inputs reaching a neuron. Thus a study of learning potential, and of the mechanisms that allow it, may also be a study of the role played by forgetting, and of its underlying mechanisms.

Despite the abundance of theory on this topic, nucleus RA of the song-control system offers the only vertebrate example to date of demonstrated synaptogenesis in relation to acquisition of a new, learned behavior.

NEUROGENESIS AND NEURONAL REPLACEMENT Let us accept, as a hypothesis, that synaptogenesis and synaptic turnover are strategies for restoring learning potential. If this hypothesis is correct, what would be the advantage of replacing neurons? I suggest, as a second hypothesis, that the DNA of neurons that partake in learning is affected by environmental factors, so that some genes are modified, or turned on or off, in an irreversible manner. If this second hypothesis is correct, then the only way to restore all degrees of freedom necessary for maximal flexibility in learning may be to replace used DNA by freshly minted DNA. In practical terms, we are probably talking about a replacement of old neurons by new neurons. This would be the ultimate in pathway rejuvenation.

We know of at least one place in the vertebrate brain, nucleus HVc of songbirds, where this second hypothesis may be confirmed, at least in that there is neuronal replacement. The fact that some forebrain nuclei (HVc) make use of neuronal replacement, while others (RA) merely make use of dendritic and synaptic changes suggest that the following hierarchy of strategies may be used by nervous systems as they try to preserve the learning potential of their limited network space: (1) synaptic modulation, (2) synaptic division, (3) synaptic replacement, (4) neuronal replacement. Whereas synaptic modulation and synaptic division are likely ways to encode memory, synaptic and neuronal turnover may determine what can be learned and what is forgotten.

What Do We Remember?

There is no evidence at present that new synapses and new neurons replace old synapses and old neurons in the human brain. If this process does occur, one might expect it in parts of the forebrain involved with the processing and storing of sensory information and perhaps motor skills. Memories lose, with time, their fine grain. Might a fine-to-coarse grain change in memories occur as the number of neurons or synapses related to a particular memory diminishes, as the particular neurons and synapses are replaced by fresh neurons and fresh synapses?

Here is a familar example, to illustrate what I have in mind. We all recognize the faces of friends and relatives as they are now, or were recently. This is a vivid memory, and small changes are noticed. How long does, or should, this memory last? How many of us remember these same faces as they were 10, 20, or 30 years ago? The image blurs. Those of us that have kept continuous series of photographs can more easily bridge the memory span. Otherwise, familiar faces from long ago have been replaced in our memory by more recent ones. This updating of memories may be a way of working around our limited memory space at little cost, because faces from the past cease to be important. Otherwise, much of our apparent remembering may be a reconstruction, so that we claim a fine-grain recollection when in reality all that is left is a coarse-grain picture updated by more recent inputs.

Hope for a New Neurology

Synaptogenesis and neurogenesis have been described here in the context of restoration of circuit plasticity for learning. Both phenomena can also be seen as a spontaneous form of brain rejuvenation or repair. At present clinical neurology relies heavily on methods that remove damaged or abnormal tissue, prevent infection, maintain electrolyte balance, and regulate ventricular pressure. Until very recently, the rest of the process of recovery of function following lesions has been left to prayer and good nurses. We are very probably at the threshold of an era when more daring measures will be attempted (Shatz, 1982). In the not-so-distant future it may be possible to treat local regions of the brain to activate genes that induce dendritic retraction and growth, that induce synaptic formation and shedding, that induce birth, migration, and differentiation of new neurons. If stem cells are not available, perhaps neuroblasts will be introduced to repair network damage.

This optimistic outlook brings me back to the issue of laterality and underused brain space. It may not be possible to rebuild brain circuits destroyed by tumors, strokes, or other lesions. It may be easier in some instances to restore lost functions by rejuvenating parts of the brain which, though underused, have lost their learning potential. Thus, even as progress is made at the cellular and molecular bases of learning, progress will also have to be made at the level of hemispheric specialization.

The opportunities that lie ahead are enormous. New insights on neuronal and hemispheric events related to learning — and forgetting — may lead to a new science of network rejuvenation and repair, a new neurology.

References

Alkon, D. L., and Crow, T. J. 1980. Associative behavioral modification in *Hermissenda:* cellular correlates. *Science* 209:412–414.
Arnold, A. P. 1980. Sexual differences in the brain. *Amer. Sci.* 68:165–173.
Baker, M. C., Spittler-Nabors, K., and Bradley, D. C. 1981. Early experience determines song dialect responsiveness of female sparrows. *Science* 214:819–821.
Baudry, M., Oliver, M., Creager, R., Wieraszko, A., and Lynch, G. 1980. Increase in glutamate receptors following repetitive electrical stimulation of hippocampal slices. *Life Sci.* 27:325–330.
Bayer, S. A., Yackel, J. W., Puri, P. S. 1982. Neurons in the rat dentate gyrus granular layer substantially increase during juvenile and adult life. *Science* 216:890–892.
Baylis, J. R. 1983. Avian vocal mimicry: its function and evolution. In D. E. Kroodsma and E. H. Miller, eds., *Acoustic Communication in Birds.* New York: Academic Press, vol. 2, pp. 51–83.
Borror, D. J., Reese, C. R. 1956. Mockingbird imitations of a Carolina wren. *Bull. Mass. Audubon Soc.* 40:245–250.
Canady, R., Kroodsma, D., and Nottebohm, F. 1981. Significant differences in volume of song control nuclei are associated with variants in song repertoire in a free-ranging songbird. *Soc. Neurosci. Abstr.* 7:845.
Carlin, R. K. and Siekevitz, P. 1983. Plasticity in the central nervous system: do synapses divide? *Proc. Natl. Acad. Sci. USA* 80:3517–21.
Changeux, J. 1974. Some biological observations relevant to a theory of learning. From Colloques Internationaux du Centre National de la Recherche Scientifique, no. 206, *Current Problems in Psycholinguistics,* pp. 281–288.
Changeux, J., Courrège, P., and Danchin, A. 1973. A theory of the epigenesis of neuronal networks by selective stabilization of synapses. *Proc. Nat. Acad. Sci. USA* 70:2974–78.
Cotman, C. W., and Nieto-Sampedro, M. 1982. Brain function, synapse renewal, and plasticity. *Ann. Rev. Psychol.* 33:371–401.
Dennis, M., and Kohn, B. 1975. Comprehension of syntax in infantile hemiplegics after cerebral hemidecortication: left hemisphere superiority. *Brain and Lang.* 2:472–482.
Dennis, M., and Whitaker, H. A. 1976. Language acquisition following

hemidecortication: linguistic superiority of the left over the right hemisphere. *Brain and Lang.* 3:404-433.
DeVoogd, T. J., and Nottebohm, F. 1981a. Gonadal hormones induce dendritic growth in the adult brain. *Science* 214:202-204.
DeVoogd, T. J., and Nottebohm, F. 1981b. Sex differences in dendritic morphology of a song control nucleus in the canary: a quantitative Golgi study. *J. Comp. Neurol.* 196:309-316.
DeVoogd, T. J., Nixdorf, B. and Nottebohm, F. 1982. Recruitment of synapses into a brain network takes extra space. *Soc. Neurosci. Abs.* 8.
Eccles, J. C. 1973. *The Understanding of the Brain.* New York: McGraw-Hill.
Fifkova, E., and Van Harreveld, A. 1977. Long-lasting morphological changes in dendritic spines of dentate granular cells following stimulation of the entorhinal area. *J. Neurocytol.* 6:211-230.
Galaburda, A. M., LeMay, M., Kemper, T. L., and Geschwind, N. 1978. Right-left asymmetries in the brain. *Science* 199:852-856.
Geschwind, N., and Levitsky, W. 1968. Human brain: left-right asymmetries in temporal speech region. *Science* 161:186-187.
Goldman, S. A., Nottebohm, F. 1983. Neuronal production, migration and differentiation in a vocal control nucleus of the adult female canary brain. *Proc. Natl. Acad. Sci. USA* 80:2390-94.
Graziadei, P. P. C., and Monti-Graziadei, G. A. 1979. Neurogenesis and neuron regeneration in the olfactory system of mammals. I. Morphological aspects of differentiation and structural organization of the olfactory sensory neurons. *J. Neurocytol.* 8:1-18.
Gurney, M. E. 1981. Hormonal control of cell form and number in the zebra finch song system. *J. Neurosci.* 1:658-673.
Gurney, M. E., and Konishi, M. 1980. Hormone-induced sexual differentiation of brain and behavior in zebra finches. *Science* 208:1380-83.
Güttinger, H. R. 1981. Self-differentiation of song organization rules by deaf canaries. *Z. Tierpsychol.* 56:323-340.
Howard, R. D. 1974. The influence of sexual selection and interspecific competition on mockingbird song. *Evolution* 28:428-438.
Immelmann, K. 1969. Song development in the zebra finch and other estrildid finches. In R. A. Hinde, ed., *Bird Vocalizations*, Cambridge: Cambridge University Press, pp. 61-74.
Kandel, E. R. 1978. *A Cell-Biological Approach to Learning.* Monograph, Society for Neuroscience, Bethesda, Maryland.
Kandel, E. R., and Schwartz, J. H. 1982. Molecular biology of learning: modulation of transmitter release. *Science* 218:433-443.
Kaplan, M. S. 1981. Neurogenesis in the 3-month old rat visual cortex. *J. Comp. Neurol.* 195:323-338.
Kaplan, M. S. and Hinds, J. W. 1977. Neurogenesis in the adult rat: electron microscopic analysis of light radioautographs. *Science* 197:1092-94.
Katz, L. C., and Gurney, M. E. 1981. There are auditory neurons in the zebra finch motor system for song. *Brain Res.* 221:192-197.

Kelley, D. B., and Nottebohm, F. 1979. Projections of a telencephalic auditory nucleus — field L — in the canary. *J. Comp. Neurol.* 183:455-470.
Kinsbourne, M., and Cook, J. 1971. Generalized and lateralized effects of concurrent verbalization on a unimanual skill. *Q. J. Exp. Psychol.* 23:341-345.
Konishi, M. 1965. The role of auditory feedback in the control of vocalization in the white-crowned sparrow. *Z. Tierpsychol.* 22:770-783.
Kroodsma, D., and Pickert, R. 1980. Environmentally dependent sensitive periods for avian vocal learning. *Nature* 288:477-479.
LeMay, M. 1976. Morphological cerebral asymmetries of modern man, fossil man and nonhuman primates. In S. R. Harnad, H. D. Steklis, and J. Lancaster, eds., *Origins and Evolution of Language and Speech*. Ann. N.Y. Acad. Sci. 280:349-366.
Liberman, A. M., Cooper, F. S., Shankweiler, D., and Studdert-Kennedy, M. 1967. Perception of the speech code. *Psychol. Rev.* 74:431-461.
McCasland, J. S. 1983. Neuronal control of bird song production. Ph. D. diss. California Institute of Technology.
McCasland, J. S., and Konishi, M. 1981. Interaction between auditory and motor activities in an avian song control nucleus. *Proc. Natl. Acad. Sci. USA* 78:7815-19.
Margoliash, D. 1983. Acoustic parameters underlying the responses of song-specific neurons in the white-crowned sparrow. *J. Neurosci.* 3:1039-57.
Marler, P. 1956. The voice of the chaffinch and its function as a language. *Ibis* 98:231-261.
Marler, P. 1970a. Birdsong and speech development: could there be parallels? *Amer. Sci.* 58:669-673.
Marler, P. 1970b. A comparative approach to vocal learning: song development in white-crowned sparrows. *J. Comp. Physiol. Psychol.* 71:25.
Marler, P., and Peters, S. 1977. Selective vocal learning in a sparrow. *Science* 198:519-521.
Marler, P., and Tamura, M. 1964. Culturally transmitted patterns of vocal behavior in sparrows. *Science* 146:1483-86.
Marler, P., and Waser, M. S. 1977. The role of auditory feedback in canary song development. *J. Comp. Physiol. Psych.* 91:8-16.
Mundinger, P. 1970. Vocal imitation and individual recognition of finch calls. *Science* 168:480-482.
Nieto-Sampedro, M., Hoff, S. F., and Cotman, C. W. 1982. Perforated postsynaptic densities: probable intermediates in synapse turnover. *Proc. Natl. Acad. Sci. USA* 79:5718-22.
Nottebohm, F. 1969a. The song of the chingolo, *Zonotrichia capensis*, in Argentina: description and evaluation of a system of dialects. *Condor* 71:299-315.
Nottebohm, F. 1969b. The "critical period" for song learning. *Ibis* 111:386-387.
Nottebohm, F. 1972. The origins of vocal learning. *Amer. Nat.* 106:116-140.

Nottebohm, F. 1975. Vocal behavior in birds. In J. R. King and D. S. Farner, eds., *Avian Biology*. New York: Academic Press, vol. 5, pp. 287–332.

Nottebohm, F. 1977. Asymmetries in neural control of vocalization in the canary. In S. Harnad, R. W. Doty, L. Goldstein, J. Jaynes, and G. Krauthamer, eds., *Lateralization in the Nervous System*. New York: Academic Press, pp. 23–44.

Nottebohm, F. 1980a. Brain pathways for vocal learning in birds: a review of the first 10 years. In J. M. S. Sprague and A. N. E. Epstein, eds., *Progress in Psychobiology and Physiological Psychology*. New York: Academic Press, vol. 9, pp. 85–124.

Nottebohm, F. 1980b. Testosterone triggers growth of brain vocal control nuclei in adult female canaries. *Brain Res*. 189:429–436.

Nottebohm, F. 1981. A brain for all seasons: cyclical anatomical changes in song control nuclei of the canary brain. *Science* 214:1368–70.

Nottebohm, F. 1984. Vocal learning and its possible relation to replaceable synapses and neurons. In D. Caplan, ed., *Biological Perspectives on Language*. Cambridge, Massachusetts: MIT Press.

Nottebohm, F., and Arnold, A. P. 1976. Sexual dimorphism in vocal control areas of the songbird brain. *Science* 194:211–213.

Nottebohm, F., and Kasparian, S. 1983. Widespread labeling of avian forebrain neurons after systemic injections of ^{3}H-thymidine in adulthood. *Soc. Neurosci. Abs*. 9.

Nottebohm, F., and Nottebohm, M. 1976. Left hypoglossal dominance in the control of canary and white-crowned sparrow song. *J. Comp. Physiol. A* 108:171–192.

Nottebohm, F., and Nottebohm, M. 1978. Relationship between song repertoire and age in the canary, *Serinus canarius*. *Z. Tierpsychol*. 46:298–305.

Nottebohm, F., Stokes, T. M., and Leonard, C. M. 1976. Central control of song in the canary, *Serinus canarius*. *J. Comp. Neurol*. 165:457–486.

Nottebohm, F., Manning, E., and Nottebohm, M. E. 1979. Reversal of hypoglossal dominance in canaries following syringeal denervation. *J. Comp. Physiol. A* 134:227–240.

Nottebohm, F., Kasparian, S., and Pandazis, C. 1981. Brain space for a learned task. *Brain Res*. 213:99–109.

Nottebohm, F., Kelley, D. B., and Paton, J. A. 1982. Connections of vocal control nuclei in the canary telencephalon. *J. Comp. Neurol*. 207:344–357.

Ojemann, G., and Mateer, C. 1979. Human language cortex: localization of memory, syntax and sequential motor-phoneme identification systems. *Science* 205:1401–3.

Okazaki-Smith, M., Chu, J., and Edmonston, W. E. 1977. Cerebral lateralization of haptic perception: interaction of responses to braille and music reveals a functional basis. *Science* 197:689–690.

Orton, S. T. 1937. *Reading, Writing and Speech Problems in Children* New York: Norton.

Paton, J. A., and Manogue, K. 1982. Bilateral interactions within the vocal control pathway of birds: two contrasting strategies. *J. Comp. Neurol.* 212:329-335.

Penfield, W., and Roberts, L. 1959. *Speech and Brain Mechanisms*. Princeton: Princeton University Press.

Price, P. H. 1979. Development determinants of structure in zebra finch song. *J. Comp. Physiol. Psych.* 93:260-277.

Ross, E. D., and Mesulam, M-M. 1979. Dominant language functions of the right hemisphere: prosody and emotional gesturing. *Arch. Neurol.* 36:144-148.

Shatz, C. J. 1982. Neural development: implications for recovery from injury. In J. G. Nicholls, ed., *Repair and Regeneration of the Nervous System*. New York: Springer-Verlag, pp. 289-311.

Sperry, R. W. 1974. Lateral specialization in the surgically separated hemispheres. In F. O. Schmitt and F. G. Worden, eds., *The Neurosciences: Third Study Program*. Cambridge, Massachusetts: MIT Press, pp. 5-19.

Sperry, R. W. 1982. Some effects of disconnecting the cerebral hemispheres. *Science* 217:1223-26.

Thorpe, W. H. 1958. The learning of song patterns by birds, with special reference to the song of the chaffinch, *Fringilla coelebs*. *Ibis* 100:535-570.

Thorpe, W. H. 1961. *Bird Song*. London: Cambridge University Press.

Chapter 8
Behavioral Asymmetry

Victor H. Denenberg

One of the exciting things about the study of laterality is that the field is inherently interdisciplinary in nature. Consider behavioral asymmetry, the topic of this chapter. The basic datum of this field was undoubtedly the observation that the vast majority of people in all cultures studied are right-handed. But that bit of information, in isolation, did not lead to scientific advances in our understanding of the phenomenon. Those occurred with the discovery that lesions on the left side of the brain, in what are now called Broca's area and Wernicke's area, resulted in language disorders often associated with inability to use the right hand. The triad of cerebral dominance, speech, and handedness emerged as a way of characterizing the human species.

The characterization was a functional one, however, since the brain was assumed to be morphologically symmetrical. Even though there were a few who suggested otherwise, over a hundred years elapsed between Broca's first paper on the lateralization of language in the early 1860s and the first quantitative demonstration of morphological asymmetry. Geschwind and Levitsky (1968) reported that the planum temporale on the upper surface of the left temporal lobe was larger than the corresponding region of the right side. They found that 65 of 100 adult brains had a larger left planum, whereas only 11 were larger on the right. These data were in the same direction as the known distributions of handedness and neural control of speech, and the authors concluded: "Our data show that this area [the planum temporale] is significantly larger on the left side, and the differences observed are easily of sufficient magnitude to be compatible with the known functional asymmetries" (Geschwind and Levitsky, 1968, p. 187).

We see here the historical interplay of behavioral and biological factors. From the observation of functional differences in handedness, we move to the structure-function correlation between left-hemisphere damage associated with impairment of speech and handedness, and finally arrive at the discovery of structural asymmetry between the left and right planum temporale. This last finding brings us back to the original functional asymmetries of handedness and speech, thus closing the circle composed of two equal arcs of behavior and biology.

Interestingly, brain-behavior asymmetries may themselves be symmetrical or asymmetrical. A symmetrical set involves observable asymmetries in both behavior and structure. An example is the apparent correlation between observed handedness and the lateralization of the neural centers controlling speech. An asymmetrical set shows laterality in either manifest behavior or structure. The best-known example is speech reception and production; despite the lack of obvious behavioral asymmetry, they are nonetheless the products of morphological asymmetries in the brain. This raises the mirror-image question of whether there are behavioral asymmetries without correlated structural asymmetries; only further empirical studies can answer this question.

It is apparent, therefore, that the study of behavioral asymmetry in isolation is of limited value in advancing our understanding of the nature of laterality. It is necessary to relate the functional asymmetry to the underlying brain morphology and/or neurochemistry. Also, as will be pointed out, the failure to find behavioral asymmetry does *not* necessarily imply a concomitant lack of neural asymmetry.

The Measurement of Asymmetry

The vast majority of behavioral measures are not designed to assess laterality. The key element necessary for a scale of laterality is a zero point so that scores below zero indicate one direction of bias (say, left) and scores above zero reflect biases in the opposite direction. An individual scoring zero would be without bias in either direction. Questionnaires to assess handedness (Annett, 1972) are a good example of this kind of scale.

The major advantage offered by such a scale is that it gives a direct measure of behavioral asymmetry for an individual with an intact brain. However, it does not specify the location in the brain of the areas associated with the lateralized behavior. In marked contrast, a behavioral scale without a laterality zero point only yields information on asymmetry if the individual has brain damage. Speech is obviously not a lateralized behavior, but its production and reception are controlled by neural centers located in the left hemisphere (at least for the vast majority of humans). Thus, speech difficulty following brain injury immediately suggests that part of the damage is in the left hemisphere.

Before presenting models of the role of laterality in brain-behavior relationships, we need to examine some aspects of brain function.

HEMISPHERIC ACTIVATION One of our constructs is that some brain regions are dominant, or specialized, for certain behavioral functions. We have already mentioned Broca's and Wernicke's areas, which are functionally specialized for different aspects of language. Similarly, the human right hemisphere is considered to be specialized for certain visuospatial functions (Geschwind 1979). The hemispheric specializations are relative rather than absolute. Thus, the right hemisphere is usually endowed with some language function. I shall use the work "activation" to indicate the extent to which an area manifests its specialized function.

INTERHEMISPHERIC INHIBITION The operational criterion which establishes that the brain is activated or specialized for a particular behavioral function is that destruction of that part of the brain results in loss of function. Thus, a stroke in certain parts of the left hemisphere will almost inevitably result in speech difficulties for right-handed individuals. In general, we say that when a brain lesion is correlated with the *disappearance* of a behavior, the lesioned area was activated or specialized for that behavior.

It is also known that there are times when a brain lesion is followed by the *appearance* of behaviors not normally observed. For example, Gainotti (1972) reported that patients with left-sided lesions were highly likely to have depressive reactions including anxiety and tears; those with right-sided lesions often manifested an "indifference reaction," such as minimizing the illness, joking about their difficulties, and being apathetic toward failure and toward events involving their families. These behaviors are not generally seen in normally functioning individuals. The appearance of new or different behavior patterns following brain injury leads to the hypothesis that the lesion brought about their occurrence, presumably by removing inhibitory control over parts of the opposite hemisphere or other portions of the same hemisphere.

A second example of an inhibitory process is seen in a study in which removal of the left neocortex of rats stimulated in infancy resulted in an increase in mouse killing (Garbanati et al., 1983). The frequency of killing by animals with intact brains was very similar to the frequency of killing by those with a right neocortical lesion, while the left-lesion group had a much higher percentage of killing. The left hemisphere apparently was acting to inhibit the mouse-killing response of the right hemisphere in animals with intact brains.

Inhibition occurs in many contexts. As an example, in the dichotic listening test different information is presented simultaneously to the two ears, yet the subject typically will report preferentially what is heard by the right ear (if the information is verbal). This has been interpreted as evidence for cross-hemispheric inhibition (Studdert-Kennedy and Shankweiler, 1970).

In our model of brain function activation and inhibition are assumed to combine algebraically, with performance being a resultant of the algebraic outcome.

We have discussed the two types of behavioral measurement and two constructs concerning brain dynamics. These can now be put together to develop some models.

Models of Brain-Behavior Relationships

Here I shall discuss models based upon measurement scales with and without a laterality zero point. The reader interested in more detailed information should consult my previous publications on the subject (Denenberg, 1980, 1981).

MEASUREMENT PROCEDURES WITH A LATERALITY ZERO POINT With a measurement scale containing a laterality zero point, in some cases the mean of a group of subjects differs significantly from zero. One may then conclude that the population from which the sample was drawn is asymmetrically biased. Regardless of whether or not the mean is significantly different from zero, it is not possible to make a determination about brain dynamics without further information. One needs data from several groups of individuals with brain damage. A minimum of two additional groups is required, each with damage in homologous regions of the brain. The means of these three groups are specified as μ_W, μ_L, μ_R, which represent respectively the population means for the group with a whole brain, for the group with only the left hemisphere intact, and for the group with only the right hemisphere intact. Two statistical tests are all that is necessary to give us information concerning the brain processes of activation and inhibition. These tests determine whether or not one can reject the following null hypotheses: $\mu_W = 0$ and $\mu_L + \mu_R = 0$.

Testing the first hypothesis shows whether the group with the intact brain differs significantly from zero. Testing the second determines whether the magnitudes of the effects of lesions in homologous regions in the two hemispheres are equal. (Normally one subtracts the two means when testing the null hypothesis. In this instance the two are added, because the scaling gives positive values to rightward responses and negative values to leftward responses. If there is no bias, the two will sum to zero.)

For each of the two hypotheses there are three possible outcomes: the difference can be found not to differ from zero, to be significant in a negative direction, or to be significant in a positive direction. These nine combinations of outcomes are shown in Fig. 8.1.

In Fig. 8.1A it is assumed that the group with an intact brain has a mean score significantly different from zero in a negative (leftward) direction. This is stated at the top of the first column and is symbolized in the second column by the two hemispheres, with the right larger than the left, based on the assumption that preferential behavior leftward is controlled by the contralateral side of the brain (in this instance the right side). The three outcomes of the test of the second statistical hypothesis are itemized in the first column of Fig. 8.1A and symbolically depicted in the second column. Here the hemispheres are shown as separated because of the surgical intervention, whereas they are coupled at the top of the column to represent the intact brain. The conclusions in the third column and the models of the intact brain in the fourth column are derived by joining the hemispheric diagrams of the second column. The inner white circles of the models in the right-hand column are the same relative size as the intact brain. The outer circles are the same relative size as the separated hemispheres in the second row. The discrepancy in size between these two configurations is shown in black and reflects the reduced response of a hemisphere when coupled, as compared to the response of that hemisphere when separated. This decreased responsiveness is assumed to be caused by inhibition of the hemisphere. The arrows show the direction of inhibition, from the inhibiting hemisphere to the inhibited one.

The brain models combine the results of the two statistical tests into one conclusion. In case 1, the group with the intact right hemisphere is assumed to have a significantly greater response bias than the group with the intact left hemisphere. This mirrors exactly what was found in the intact group and leads to the conclusion that greater right-hemispheric activation accounts for the behavior of the group with an intact brain. In case 2, we assume that the two lesion groups do not differ significantly, yet the intact group does differ in a negative direction. Since the two lesion groups did not differ, there is no evidence that either hemisphere is more activated or specialized than the other, so the construct of activation cannot be used to explain the behavior of the intact group (as in case 1). Instead, we conclude that the difference seen in the intact group is a consequence of the right hemisphere's inhibition of the left. In case 3, we have the odd finding that the left-hemisphere group has a greater response bias than does the right group. The significant difference between the two lesion groups is indicative of an activation difference, but since that difference is in the opposite direction from that of the intact group, we must also introduce the construct of inhibition. Our conclusion is that in the intact animal the left hemisphere is more activated than the right by the test situation, but that the right hemisphere overrides the left's activation by very strong inhibition.

When the mean of the intact group is positive and significant, the relationships shown in Fig. 8.1A are mirror-reversed. These are shown in Fig. 8.1B.

The last set of outcomes is generated when the hypothesis that $\mu_W = 0$ cannot be rejected. These are shown in Fig. 8.1C. It is important to note that the failure to find behavioral asymmetry does not mean that there is no neural asymmetry. In cases 7 and 9 one has evidence of activation differences in the two hemispheres that are counteracted by cross-hemispheric inhibition. Only if case 8 obtains can one conclude that the lack of behavioral asymmetry is correlated with the lack of neural asymmetry.

MEASUREMENT PROCEDURES WITHOUT A LATERALITY ZERO POINT

When a scale has a laterality zero point we can make a statistical test on the performance of the intact group, W, but without such a scale this is not possible. To compensate, it is necessary to have three statistical tests, rather than two as discussed previously. The three null hypotheses are $\mu_L = \mu_R$, $\mu_W = \mu_L$, and $\mu_W = \mu_R$.

Figure 8.2 summarizes the outcome of these tests. As before, we assume that the performance of the intact group reflects the relative performance of each hemisphere as a resultant of activation and inhibition combined.

Let us now apply these brain-behavior models to a series of research studies on laterality in rats.

Fig. 8.1 Models of brain for measurements that have a laterality zero point. RH = right hemisphere; LH = left hemisphere; μ_W = mean score of group with whole brain intact; μ_L mean of group with only LH intact; μ_R = mean of group with only RH intact. In right-hand column, arrow points to inhibited hemisphere.

In A, $\mu_W < 0$. The second column indicates that in the intact brain performance is dominated by RH; that when the hemispheres are separated, there are three cases — RH performance > LH performance (case 1); RH performance = LH performance (case 2); RH performance < LH performance (case 3).

In B, $\mu_W > 0$. The second column indicates that in the intact brain performance is dominated by LH; when the hemispheres are separated, there are three cases — LH performance > RH performance (case 4); LH performance = RH performance (case 5); LH performance < RH performance (case 6).

In C, $\mu_W = 0$. The second column indicates that in the intact brain LH performance = RH performance; when the hemispheres are separated, there are three cases — LH performance < RH performance (case 7); LH performance = RH performance (case 8); LH performance > RH performance (case 9).

(All three figures after Denenberg, 1980; by permission of the American Physiological Society and the American Journal of Physiology.)

Fig. 8.1A

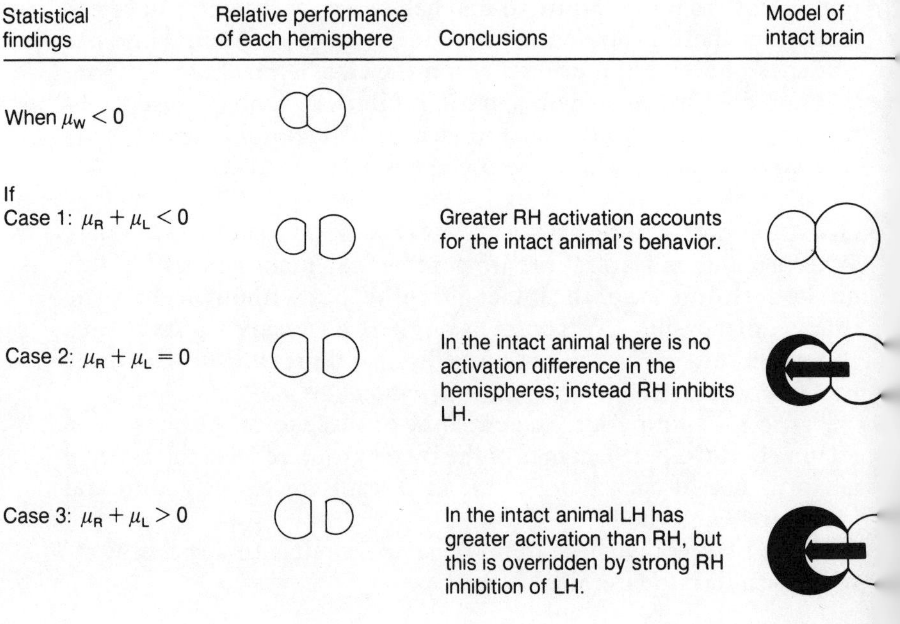

Statistical findings	Relative performance of each hemisphere	Conclusions	Model of intact brain
When $\mu_W < 0$			
If Case 1: $\mu_R + \mu_L < 0$		Greater RH activation accounts for the intact animal's behavior.	
Case 2: $\mu_R + \mu_L = 0$		In the intact animal there is no activation difference in the hemispheres; instead RH inhibits LH.	
Case 3: $\mu_R + \mu_L > 0$		In the intact animal LH has greater activation than RH, but this is overridden by strong RH inhibition of LH.	

Behavioral Asymmetry   121

Fig. 8.1B

Statistical findings	Relative performance of each hemisphere	Conclusions	Model of intact brain
When $\mu_w > 0$	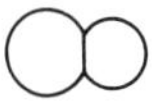		
If Case 4: $\mu_R + \mu_L > 0$		Greater LH activation accounts for the intact animal's behavior.	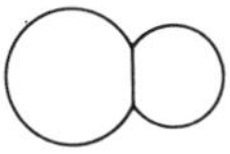
Case 5: $\mu_R + \mu_L = 0$		In the intact animal there is no activation difference in the hemispheres; instead LH inhibits RH.	
Case 6: $\mu_R + \mu_L < 0$		In the intact animal RH has greater activation than LH, but this is overridden by strong LH inhibition of RH.	

Fig. 8.1C

Statistical findings	Relative performance of each hemisphere	Conclusions	Model of intact brain
When $\mu_w = 0$			
Case 7: $\mu_R + \mu_L < 0$		In the intact animal RH has greater activation than LH, but this is overridden by strong LH inhibition of RH.	
Case 8: $\mu_R + \mu_L = 0$		There is no evidence of laterality in the intact animal.	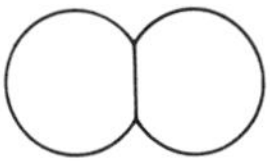
Case 9: $\mu_R + \mu_L > 0$		In the intact animal LH has greater activation than RH, but this is overridden by strong RH inhibition of LH.	

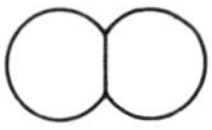

Brain-Behavior Asymmetry and the Effects of Early Experience

Research over the past 25 years has established that a form of stimulation called handling, administered to rats in infancy, results in an adult animal that is less emotional and more exploratory (Denenberg, 1964, 1969). Emotional behavior is known to be lateralized in the human brain (Gainotti, 1972) and exploratory behavior involves spatial processing, which is also lateralized in the human (Geschwind, 1979). These findings suggested that the rat's brain might be lateralized for these and other behavioral functions, and that early experiences might act to modify this laterality. Our group carried out a series of experimental studies to test these ideas. It will be convenient to present first our rearing and early experience procedures, followed by our surgical intervention techniques, and finally our findings.

EARLY REARING AND HANDLING At birth whole litters of rats are assigned to either a control (nonhandled) condition or to handling stimulation (Denenberg, 1977). On day 1 of life all litters are culled to 8 pups. The control pups are returned to their mothers and are not disturbed thereafter until weaning. The experimental pups receive handling stimulation, which consists of removing the pups from the home cage, placing each separately into a can containing shavings, leaving them there for 3 min, then returning them to the home cage. This procedure is repeated daily from day 1 through day 20. All litters are weaned when 21 days old.

After weaning, the rats are reared either in standard laboratory cages or in enriched environments. Here I shall discuss only the findings on those reared in laboratory cages. That is, the discussion is restricted to animals that were handled or nonhandled in infancy and maintained under the same conditions thereafter.

SURGICAL PROCEDURES When adult, 4 male littermates are randomly assigned to one of four surgical procedures: a left or right neocortical ablation, sham surgery, or no surgery. The intended area of ablation includes virtually all of the neocortex from pole to pole in an anterior-posterior direction, and from the sagittal to the rhinal fissure mediolaterally. In the studies to be described, no difference was found between the sham and the no-surgery groups, and their data were pooled.

OPEN-FIELD ACTIVITY The open-field test measures the two behavioral dimensions of emotionality and exploration (Whimbey and

Denenberg, 1967), and these were the variables we studied in our first experiment (Denenberg et al., 1979). After recovery from brain surgery, handled and nonhandled animals were tested for 4 days in the field and their activity recorded. The data are presented in Table 8.1. Removing either hemisphere increased the activity of nonhandled animals, but the two lesion groups did not differ from each other. The handled group did show asymmetry, with the right-lesion animals (left neocortex intact) having more than twice the activity of those with a left lesion.

Since open-field activity does not have a laterality zero point, we turn to the models in Fig. 8.2. Case 11 depicts the data of the nonhandled rats: there is no evidence of laterality from the two lesion groups; since removing either hemisphere increased activity, we conclude that both hemispheres are activated by the open-field task and each also inhibits the other. For the handled subjects, case 17 applies, that is, in this test situation the left hemisphere has greater activation than the right, but at the same time is inhibited by the right hemisphere.

Table 8.1 Effects of brain lesions on behavior of nonhandled (NH) and handled (H) rats.

Dependent variable[a]	Whole brain	Right lesion	Left lesion
Open-field activity			
NH	8.90[b]	22.33	27.64
H	12.51	36.27[b]	17.91
Left-right spatial choice			
NH	0.030	0.508[c]	−1.027[c]
H	−0.386[d]	0.330	−0.739
Taste aversion			
NH	23.9	23.8	23.7
H	28.7[e]	25.6[e]	21.7[e]
Muricide			
NH	96.0[b]	68.8	75.0
H	78.0	67.6	94.6[b]

a. Measurement units: activity — number of squares entered; spatial choice — see text; taste aversion — ml milk ingested; muricide — percentage of animals that killed. Statistical tests used: t-test for open-field activity and taste aversion; chi-square for muricide.
b. Differs significantly from other values in row.
c. Differ significantly from each other.
d. Differs significantly from zero.
e. All differences are significant, that is, $p \leq .05$.

Fig. 8.2 *Models of brain for measurements without a laterality zero point. Symbols are as in Fig. 8.1. (After Denenberg, 1980; by permission of the* American Physiological Society *and the* American Journal of Physiology.)

Statistical findings	Conclusions	Model of intact brain
Given that $\mu_L = \mu_R$		
If Case 10: $\mu_W = \mu_L$ and $\mu_W = \mu_R$	There is no evidence of laterality.	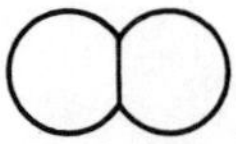
Case 11: $\mu_W < \mu_L$ and $\mu_W < \mu_R$	There is no evidence of laterality; both hemispheres are activated, and each inhibits the other.	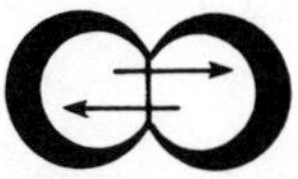
Given that $\mu_L < \mu_R$		
If Case 12: $\mu_W = \mu_R$ and $\mu_W > \mu_L$	Greater RH activation accounts for the behavior.	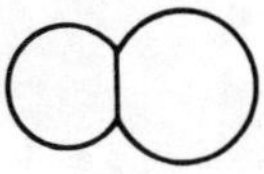
Case 13: $\mu_W < \mu_R$ and $\mu_W = \mu_L$	RH has greater activation than LH, but this is overridden by LH inhibition of RH.	
Case 14: $\mu_W < \mu_R$ and $\mu_W > \mu_L$	RH has greater activation than LH, but RH is partially inhibited by LH.	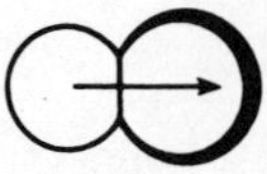

Statistical findings	Conclusions	Model of intact brain
Case 15: $\mu_W < \mu_R$ and $\mu_W < \mu_L$	Both RH and LH are activated, RH to a greater extent; but each also inhibits the other.	

Given that $\mu_L > \mu_R$

If Case 16: $\mu_W = \mu_L$ and $\mu_W > \mu_R$	Greater LH activation accounts for the behavior.	
Case 17: $\mu_W < \mu_L$ and $\mu_W = \mu_R$	LH has greater activation than RH, but is overridden by RH inhibition of LH.	
Case 18: $\mu_W < \mu_L$ and $\mu_W > \mu_R$	LH has greater activation than RH, but RH is partially inhibited by LH.	
Case 19: $\mu_W < \mu_L$ and $\mu_W < \mu_R$	Both LH and RH are activated, LH to a greater extent; but each also inhibits the other.	

LEFT-RIGHT SPATIAL CHOICE   The open field measures both emotionality and exploration. In our next three studies we tried to evaluate these dimensions separately. To measure spatial preference (a facet of exploration) we placed rats in one corner of an open field and observed whether they went left or right after leaving the starting square (Sherman et al., 1980). They were given four tests, and a laterality index was devised by the formula $(R - L)/(R + L)^{1/2}$, where R = number of right choices and L = number of left choices. This scale has a laterality zero point, with negative scores reflecting a leftward bias and positive scores a rightward bias. The findings are summarized in Table 8.1.

Nonhandled rats with intact brains had no directionality bias, but the two nonhandled lesion groups did differ, with the left lesion group having the greater magnitude of response. These data fit case 7 (Fig. 8.1C). We conclude that in the left-right spatial-choice situation the normal control rat has a more activated right hemisphere that is counterbalanced by an inhibiting process from the left hemisphere, with the result that the intact animal does not appear to be lateralized.

The situation is somewhat different for the handled group. Here the intact group had a significant leftward bias. The difference between the two handled lesion groups was −0.409, which is greater than the whole brain's mean of −0.386. Even though the difference was not significant, other data (discussed in Denenberg, 1980) led to the conclusion that case 1 was the most appropriate model. We see, then, for both handled and nonhandled animals that the right hemisphere is more activated when making a left-right spatial choice, inclining the animal to go to the left. This bias is inhibited in the nonhandled animals, but is fully expressed in those handled in infancy.

TASTE AVERSION   To study an emotional process independent of any spatial bias, we used a procedure called the taste aversion paradigm to induce a learned fear. We first induced the fear, next lesioned the brains, then studied the retention of the learned fear (Denenberg et al., 1980).

To induce the learned fear we had the rats drink a sweetened milk solution and followed this by an injection of lithium chloride, causing a major gastric disturbance. The animal associated the unique taste of the milk with the stomach upset and in this way learned to avoid milk. Three weeks after this experience the animals underwent surgery, and a month later were again exposed to the sweetened milk for a half-hour daily for 13 days. The data in Table 8.1 are the mean amounts ingested during the last 6 days of testing.

There were no differences among the three nonhandled groups. All the handled groups differed from one another. Those with a remaining intact right hemisphere consumed the least milk, thereby indicating the greatest retention of fear, followed by those with an intact left hemisphere; those with a fully intact brain showed the least fear in that they consumed the greatest amount of milk. Case 15 in Fig. 8.2 is a model of these data. The taste aversion procedure activated fear in both hemispheres, but to a greater extent in the right; in addition, each hemisphere acted to inhibit the fear response of the other so that those with intact brains appeared to have the last fear.

MURICIDE   In the taste aversion experiment, which involved a learned emotional response, we found that the right hemisphere was more strongly involved than the left. In our next study we examined muricide, or mouse killing, since this is a spontaneously occurring emotional behavior (Garbanati et al., 1983). Rats, handled or undisturbed in infancy, had the usual surgical procedures, and several months later were tested for mouse killing. The presence or absence of killing was recorded. The data in Table 8.1 reveal no evidence of asymmetry in the nonhandled groups, although the lesion reduced the incidence of killing. There is strong evidence of asymmetry in the handled groups: rats having an intact right hemisphere had a greater killing frequency than those with only an intact left hemisphere. On the assumption that a higher frequency of killing reflects greater fear, case 13 models the findings: the right hemisphere is more activated to kill than is the left, but in the intact animal this activation is overridden by an inhibitory process from the left hemisphere.

A TEST OF THE INHIBITION MODEL   If the conclusion reached above is correct, it follows that severing the connections between the hemispheres should eliminate the inhibitory process that attenuates the mouse-killing response. The corpus callosum is the largest fiber bundle connecting the hemispheres and is therefore the most likely candidate for transmitting the inhibitory impulses. We have recently tested this hypothesis and obtained confirmatory data (Denenberg et al., 1983).

We first repeated the muricide experiment described above. The findings are shown in Table 8.2. These data are in substantial agreement with the original study, confirming that animals stimulated in

Table 8.2 Percentages of rats that killed mice as a function of early stimulation and brain lesion.

Treatment in infancy	Whole brain	Right lesion	Left lesion
Nonhandled	63.6	40.0	58.8
Handled	34.8	43.8	76.5[a]

a. Differs significantly from whole brain group, and from right lesion group at 0.07 level (one-tailed chi-square test).

infancy have lateralized brains for muricide with the right-hemisphere-intact group having the greater killing response. Having established replicability, we then examined the effects of severing the corpus callosum. We compared handled rats with intact brains to others which had the corpus callosum split. Within each of these conditions were two subgroups: some rats had had early social exposure to mice while others had not. The findings are summarized below.

	Intact brain	Split brain
Exposed to mice	34.8%	66.7%
Not exposed	31.6%	61.5%

When the two intact groups are pooled (mean = 33.3% killers) and the two split groups are pooled (mean = 63.6%), the splits differ significantly from the intacts and have a mean value quite similar to those with a left neocortical lesion in Table 8.2. The data do confirm the cross-hemisphere inhibition hypothesis, but additional studies are needed to establish the replicability and generality of this finding.

Postural Asymmetry, Sex Differences, and Testosterone

NEONATAL POSTURAL ASYMMETRY An important advance in the study of behavioral asymmetry occurred when Ross et al. (1981) demonstrated that the neonatal rat has postural asymmetry. This showed that asymmetry was present from birth and opened up the possibility of developmental studies. Their procedure was as follows. Pups were placed into a bilaterally symmetrical position; upon their release they assumed an asymmetrical posture. Tail asymmetry was particularly obvious and was used as the index of postural asymmetry. Females had on the average a right tail bias; males did not differ from chance. These findings paralleled the previous report of these researchers that the adult Sprague-Dawley female is biased to turn rightward (Glick and Ross, 1981). They also found that in 28 of 33 animals the direction of rotation in adulthood was the same as their neonatal tail direction, thus establishing a strong association between postural asymmetry in infancy and spatial bias in adulthood.

We used Purdue-Wistar rats, having previously determined that these animals are on the average left-biased in their spatial preference (Sherman et al., 1980). To generalize from the findings of Ross et al. (1981), we would predict that rats from our population should exhibit a left-tail bias in infancy. This hypothesis has been tested and confirmed (Denenberg et al., 1982). The data on the numbers and percentages of animals in each group were as follows:

	Left-tail bias	Neutral	Right-tail bias
Male	591 (51.9%)	130 (11.4%)	418 (36.7%)
Female	626 (57.7%)	63 (5.8%)	395 (36.4%)

Both male and female Wistar rats are left biased, the females significantly more so than the males. These findings (1) confirm the results of the Ross group and establish their generality; (2) demonstrate that there are strain differences, the Sprague-Dawley being right biased while the Wistar rat is left biased; and (3) document sex differences, with the female being more asymmetrical in both rat populations.

PRENATAL TESTOSTERONE AND TAIL POSTURE ASYMMETRY The sex difference in tail posture suggested that there may be a hormonal basis for this behavior. If so, then modification of hormonal levels during fetal life could modify this response, perhaps even reverse it. To investigate this, pregnant rats were injected during the last trimester with testosterone propionate (TP), dihydrotestosterone propionate (DHTP), sesame oil, or nothing. Tail posture was assessed the day after birth. The sesame oil and no-injection groups did not differ, and their data were pooled. The findings are summarized in Table 8.3.

Control males and females had the same patterns as found previously: both groups were biased to the left, with the females more so. Within the female groups, controls and pups whose mothers had received DHTP in utero did not differ from each other, but both differed from pups whose mothers had received TP in utero. In contrast to the controls and the DHTP group, the females exposed to testosterone showed a shift of neonatal tail posture so that the majority now had a rightward bias. There were no differences among the three male groups.

Although TP can be converted to estradiol in certain locations in the brain, DHTP cannot. TP actions on the brain often take place in sites of such conversion, although DHTP does sometimes directly affect the central nervous system (CNS). Thus, the effectiveness of TP in shifting the female's tail posture distribution rightward, and the inability of DHTP to do so, is taken to suggest that TP is exerting its effect on tail posture via a CNS site in which conversion takes place. The lack of differences among the male groups is consistent with other data in the literature suggesting that at least in the rat the female, but not the male, appears to be a more sensitive animal model to evaluate many of the prenatal effects of androgen. One important implication of these data is that testosterone and other steroids may play a significant role in the development of behavioral asymmetry.

The findings take on added theoretical significance because they appear similar, in principle, to a study in humans by Geschwind and Behan (1982). These researchers reported that strongly left-handed individuals were more likely to have learning disabilities and immune diseases than strongly right-handed persons. These effects, they proposed, might be caused by an excessive production of (or sensitivity to) testosterone in the fetus, which results in diminution

Table 8.3 Tail posture as a function of prenatal hormone treatment. (From Rosen et al., 1983; with permission of Developmental Brain Research and Elsevier Biomedical Press.)

Treatment group	Tail posture (%)			N
	Left	Neutral	Right	
Females				
Controls	64.4	5.0	30.6	317
DHTP	60.8	4.2	35.1	148
TP	38.5	9.6	51.9	104
Males				
Controls	56.3	9.3	34.3	323
DHTP	45.8	9.0	45.2	155
TP	52.5	10.2	37.3	118

of the rate of growth of the left hemisphere and suppression of the development of the thymus gland. These two factors can account for the higher frequency of left-handedness and learning disabilities in males and for the increase in immune diseases in left-handers. The hypothesis is discussed further in Chapter 14.

We have found, in female rat pups, that excessive prenatal testosterone causes a shift in tail posture, which may be analogous to handedness in humans. If the parallel holds, then these females should also be more susceptible to immune diseases. Experiments to test this hypothesis are under way.

Two sets of experiments have been described. In the first set, some rats were given extra stimulation in infancy by a procedure called handling, while others were undisturbed controls. When mature, the rats underwent a left or right neocortical ablation, or their brains remained intact. In different experiments handled and nonhandled rats were tested for open-field performance (which measures the behavioral dimensions of emotional reactivity and exploration), left-right preference (a facet of spatial functioning), taste aversion (a learned fear), and muricide (which reflects heightened emotionality).

In the second set of experiments the postural asymmetry of neonatal rats was studied, using tail posture as the index of laterality. The influence of prenatal testosterone on this response was also investigated.

Several conclusions may be drawn from these studies:

(1) The brain of the rat is lateralized for behavioral processes. When combined with evidence of lateralization in other species (see other chapters in this volume), one may conclude that brain-behavioral lateralization is a rather general property of vertebrates.

(2) The behavioral processes lateralized in the rat to date are remarkably like those found in the human, namely, the right hemisphere is centrally involved in certain strong emotional behaviors and in some aspects of spatial function.

(3) Extra stimulation in infancy acts to enhance already-present laterality, or causes laterality to occur in situations where it is not normally seen in rats raised under usual laboratory conditions.

(4) A model of brain function has been proposed, based upon the concepts of hemispheric activation and cross-hemispheric inhibition. In the single experiment that has tested this model to date severing the corpus callosum resulted in elimination of inhibitory processes, but replication is needed to establish this finding firmly. The data do confirm the heuristic value of the model.

(5) Prenatal testosterone administration causes a shift of postural asymmetry in female rat pups, but not in males, suggesting that steroids may influence lateralization. The rat data appear compatible with previous findings in humans that have suggested that testosterone plays a role in the determination of handedness.

References

Annett, M. 1972. The distribution of manual asymmetry. *Brit. J. Psychology* 63:343–358.
Denenberg, V. H. 1964. Critical periods, stimulus input, and emotional reactivity: a theory of infantile stimulation. *Psychol. Rev.* 71:335–351.
Denenberg, V. H. 1969. The effects of early experience. In E. S. E. Hafez, ed., *The Behaviour of Domestic Animals*. London: Bailliere, Tindall & Cassell, pp. 96–130.
Denenberg, V. H. 1977. Assessing the effects of early experience. In R. D. Meyers, ed., *Methods in Psychobiology*. New York: Academic Press, vol. 3, pp. 127–147.
Denenberg, V. H. 1980. General systems theory, brain organization, and early experiences. *Am. J. Physiol. Regulatory Integrative Comp. Physiol.* 238:R3-R13.
Denenberg, V. H. 1981. Hemispheric laterality in animals and the effects of early experience. *Behav. Brain Sci.* 4:1–49.
Denenberg, V. H., Garbanati, J., Sherman, G., Yutzey, D. A., and Kaplan, R. 1979. Infantile stimulation induces brain lateralization in rats. *Science* 205:707–710.
Denenberg, V. H., Hofmann, M., Garbanati, J. A., Sherman, G. F., Rosen, G. D., and Yutzey, D. A. 1980. Handling in infancy, taste aversion, and brain laterality in rats. *Brain Res.* 20:123–133.
Denenberg, V. H., Rosen, G. D., Hofmann, M., Gall, J., Stockler, J. and Yutzey, D. A. 1982. Neonatal postural asymmetry and sex differences in the rat. *Develop. Brain Res.* 2:417–419.
Denenberg, V. H., Gall, J., Berrebi, A., Hofmann, M. B., and Yutzey, D. A. 1983. Lateralization of muricide in rats. In preparation.
Gainotti, G. 1972. Emotional behavior and hemispheric side of the lesion. *Cortex* 8:41–55.
Garbanati, J. A., Sherman, G. F., Rosen, G. D., Hofmann, M., Yutzey, D. A., and Denenberg, V. H. 1983. Handling in infancy, brain laterality and muricide in rats. *Behav. Brain Res.* 7:351–359.
Geschwind, N. 1979. Specializations of the human brain. *Sci. Amer.* 241:180–199.
Geschwind, N., and Behan, P. 1982. Left-handedness: association with immune disease, migraine, and developmental learning disorders. *Proc. Natl. Acad. Sci. USA* 79:5097–5100.

Geschwind, N., and Levitsky, W. 1968. Human brain: left-right asymmetries in temporal speech region. *Science* 161:186–187.

Glick, S. D., and Ross, D. A. 1981. Right-sided population bias and lateralization of activity in normal rats. *Brain Res.* 205:222–225.

Rosen, G. D., Berrebi, A. S., Yutzey, D. A., and Denenberg, V. H. 1983. Prenatal testosterone causes shift of asymmetry in neonatal tail posture of the rat. *Develop. Brain Res.:* 9:99–101.

Ross, D. A., Glick, S. D., and Meibach, R. C. 1981. Sexually dimorphic brain and behavioral asymmetries in the neonatal rat. *Proc. Nat. Acad. Sci. USA* 78:1958–61.

Sherman, G. F., Garbanati, J. A., Rosen, G. D., Yutzey, D. A., and Denenberg, V. H. 1980. Brain and behavioral asymmetries for spatial preference in rats. *Brain Res.* 192:61–67.

Studdert-Kennedy, M., and Shankweiler, D. 1970. Hemispheric specialization for speech perception. *J. Acoust. Soc. Amer.* 48:579–594.

Whimbey, A. E., and Denenberg, V. H. 1967. Two independent behavioral dimensions in open-field performance. *J. Comp. Physiol. Psychol.* 63:500–504.

Chapter 9

Age, Sex, and Environmental Influences

Marian Cleeves Diamond

Recent evidence indicates that the rat forebrain is more structurally asymmetrical than was previously believed. We must now refine our thinking and consider the many variations related to species, sex, age, as well as regional differences, to name a few. Once we become familiar with these factors, we will need to determine how both external and internal environments can modify innate patterns.

For almost 20 years our group has been concerned with the cerebral cortex and its alteration by the environment. We pooled anatomical data from the two hemispheres, assuming that they were similar. Only within the past few years have we examined the hemispheres separately. This chapter will review our data, bringing together new information and some that has been previously published (Diamond et al., 1971, 1972, 1975, 1979, 1981, 1982, 1983; Diamond, 1976; Dowling et al., 1982). We shall consider right-left differences in cortex, and in some cases hippocampus, (1) during development and aging in male and female Long-Evans rats; (2) in intact and gonadectomized Long-Evans male rats; (3) in intact and ovariectomized Long-Evans female rats; (4) in S_1 males from three separate environmental conditions (enriched, standard colony, and impoverished) and four different age groups.

Many questions are suggested by these studies: Do hemispheric differences remain constant during development and aging? Are sex differences also found in subcortical structures? Are right-left differences uniform in each lobe or region—for example, is the entire occipital region larger on one side?

Although we cannot yet answer all these questions, we can offer some speculations in relation to other queries: What might be the advantages of the asymmetrical patterns in the male rat, and what in turn are the advantages of the quite different asymmetrical patterns in the female? Why do some cortical areas react to sex steroid hormones whereas others do not? Why do some cortical regions respond to altered environmental conditions in similar fashion on both sides, while regions adjacent to them respond in a different fashion?

The methods employed have been published previously (Diamond et al., 1972, 1983; Diamond, 1976).

Right-Left Cortical Differences

DEVELOPMENT AND AGING IN MALE AND FEMALE RATS We first noted cortical asymmetry in the Long-Evans male rat in a developmental and aging study (Diamond et al., 1975). Using rats 6, 10, 14, 20, 41, 55, 77, 90, 185, 300, 400, and 650 days of age, we observed, in coronal sections, that the right cerebral cortex was thicker than the left in 92 out of 98 locations, ranging from anterior to posterior and from medial to lateral. An interesting pattern of asymmetry was clearly exhibited when the cortex was divided into three roughly equal divisions from medial to lateral. The middle third consistently demonstrated statistically more significant right-left differences than did the medial and lateral thirds. Cortical thickness was thus greater in the right hemisphere of the male rat in the middle third strip from anterior to posterior. Asymmetrical skull growth might conceivably determine this pattern, but one would expect skull formation to be uniformly asymmetrical only in response to internal brain growth or muscle use.

We have recently added right-left data from 900-day-old rats. Rats in previous studies lived only to about 650 days of age, but with increased effort we achieved greater longevity. In the 900-day-old male rats the right hemisphere was again thicker than the left in every area but one, area 39. The hemispheric differences were smaller than in younger animals, and none were statistically significant.

Diamond et al. (1983) examined male right-left differences by means of analysis of variance for selected age groups (6, 14, 20, 90, 185, and 400 days), for comparison with female brains from similar age groups. The data in Table 9.1 reflect changes in laterality with age and by sex.

In Table 9.1, 31 of 49 measurements in males are statistically significant, in 16 regions at the $p < 0.001$ level (cortical areas de-

Table 9.1 Statistical significance of differences between right and left cortical thickness in Long-Evans male and female rats.[a]

	Age (days)	N	Cortical area							
			10	4	3	2	18	17	18a	
Males:	6	15	0.001	0.05	0.001	N.S.	0.01	0.001	0.02	
	14	17	0.001	0.02	0.001	N.S.	0.001	0.001	0.001	
	20	15	0.001	0.01	0.001	N.S.	0.001	0.001	N.S.	
	90	15	N.S.	0.05	0.001	N.S.	N.S.	0.001	N.S.	
	185	15	0.001	N.S.	0.01	0.05	0.02	0.001	0.05	
	400	15	0.01	N.S.	0.01	0.01	0.01	0.01	N.S.	
	900	8	N.S.	N.S.	N.S.	N.S.	N.S.	N.S.	N.S.	
			M10[b]	L10[b]	3	2	18	17	18a	39
Females:	7	14–15	N.S.	N.S.	−N.S.[c]	−N.S.	N.S.	−N.S.	−N.S.	−N.S.
	14	14–17	N.S.	−N.S.	N.S.	N.S.	N.S.	N.S.	N.S.	−N.S.
	21	10–19	N.S.	N.S.	N.S.	0.01	N.S.	N.S.	N.S.	0.05
	90	19–20	N.S.	=	N.S.	−N.S.	=	N.S.	N.S.	N.S.
	180	11	−N.S.	N.S.	−N.S.	N.S.	N.S.	−N.S.	−N.S.	−N.S.
	390	17	N.S.	N.S.	0.01	N.S.	N.S.	−N.S.	−N.S.	−0.05

a. Except as noted, right-sided areas were larger than left in males.
b. Medial 10 and lateral 10.
c. A minus sign (−) indicates that the right side was larger than the left; = means no difference in size; in all other cases in females the left side was larger than the right.

fined according to Krieg, 1946). Area 2 in males is unusual in showing nonsignificant differences until 185 days of age, when a 3% ($p < 0.05$) right greater than left difference is found. A 3% difference is also seen at 400 days, but the significance level drops ($p < 0.01$). This is the only area to show a consistently increasing degree of laterality with age (except at 900 days, when significant side differences are lost in all areas). On the other hand, area 3, adjacent to 2, begins at 6 days of age with a 7% ($p < 0.001$) right greater than left difference, and by 400 days it shows only a 3% ($p < 0.01$) difference. Still another area, 17, shows the strongest laterality of any region measured, maintaining a difference of 7% ($p < 0.001$) from 6 to 185 days with a slight decrease, 6% ($p < 0.01$), at 400 days.

Because of the stability of the findings in area 17 in the rat, Murphy (1983) measured the comparable area in the human brain. In 31 brains from the Yakovlev collection, Murphy and his colleagues (unpublished data) found the volume of area 17 to be larger on the right than the left ($p < 0.05$). Rat data can therefore provide guidelines for study of laterality patterns in the human. Laterality in the male rat cortex changes with age and with area, even when the animals are housed in similar environments, but no similar data are available for humans.

Consider now right-left differences in the cortex of the female Long-Evans rat (Table 9.1). Of 54 measurements 50 show a nonsignificant right-left difference, in contrast to only 11 of 42 measurements at the same time periods in the male rat. The cortex of the male Long-Evans rat is thus more strongly lateralized than that of the female. In 36 of the 54 female measurements, the left cortex is thicker than the right. However, the strong, consistent age changes seen in males are lacking in females, except in occipital or posterior cortex. In the 7-day female the right occipital cortex is thicker than the left, but the left is thicker at 14, 21, and 90 days. In older females, 180 and 390 days of age, right occipital cortex is again thicker than left. However, none of the differences are statistically significant. Since relatively large numbers of rats were used for these studies, the data suggest a tendency for the female right occipital cortex to alter its laterality pattern during development and aging.

It has been reported that male rats are superior to females in visuospatial ability and that spatial laterality may be important for territoriality in the male (Sherman et al., 1980; Stokes and McIntyre, 1981). Right structural dominance in the visuospatial region of the cortex fits these male roles. One might offer the following hypothesis for female left dominance. The female is not so strong as the male, and therefore vocalization may be an important means of protection. With language and song predominantly in the left brain of man and birds respectively, perhaps "language" (vocalization) is also localized on the left in the female rat. Furthermore, as the young scramble from the nest, the mother needs to communicate vocally to indicate her position. Pronounced asymmetry might not be beneficial to the rat mother, of whom rapid response, both physical and mental, is required for protection of the young.

INTACT VERSUS GONADECTOMIZED MALE RATS We had no reason to expect the male hormone, testosterone, to alter cortical morphology, because we had previously studied the cortical thickness of animals living in enriched or impoverished external environments with and without gonads. In those experiments the gonads were removed either at birth or at 30 days of age, and at 60 to 90 days the animals were placed in their respective external environmental conditions (Diamond, 1976).Cortical thickness changes induced by environmental input were the same whether the testes were intact or not. At that time the data from the right and left cortices were pooled, and we did not look at hemispheric differences separately. That experiment led us to believe that perhaps testosterone did not alter cortical structure.

Later, after we found that removal of the ovaries could alter the structure of the female cortex (Diamond et al., 1979), we decided to examine more carefully the effects of gonadectomy on the male cortex. Under cryogenic anesthesia we removed the testes from Long-Evans rats at 1 day of age. The animals were returned to their mothers until weaning at 21 to 23 days, at which time they were separated, 3 to a cage (32 × 20 × 20 cm), until autopsy of all animals at 90 days of age. Intact littermates lived in similar conditions until they also reached 90 days, at which time all animals were anesthetized, perfused, and the brains removed for histological study.

In the intact Long-Evans male animals the right cortex was thicker than the left in all 7 areas measured, but statistical significance was reached in only 3 of the 7 regions (areas 4, 3 and 17). However, in the rats without testes for 89 days, 6 out of 9 regions were thicker on the *left* than on the right — a pattern similar to that of the female. The occipital cortex, however, remains right-dominant in males with or without testes, in different strains of rats, and in females without ovarian hormones (to be discussed in the next section) or with reduced ovarian hormones in later life.

Table 9.2 Differences in right and left cortical thickness between intact and gonadectomized male Long-Evans rats at 90 days of age.

Subjects	N	Cortical area						
		10	4	3	2	18	17	18a
Intact animals	15							
Laterality		R > L	R > L	R > L	R > L	R > L	R > L	R > L
Significance		N.S.	0.05	0.001	N.S.	N.S.	0.001	N.S.
Gonadectomized animals	21							
Laterality		L > R	L > R	L > R	L > R	L > R	R > L	R > L
Significance		N.S.	N.S.	0.01	0.05	N.S	0.05	0.01

In a previous publication (Diamond et al., 1981) we mentioned that gonadectomy in males did not alter the right-left hemispheric pattern at 90 days of age except in 3 regions, lateral 10, 2, and 18a. Table 9.2 shows the left-right differences in another larger group of animals with and without gonads.

INTACT VERSUS OVARIECTOMIZED FEMALE RATS Our first clue that ovarian hormones might influence the dimensions of the cortex appeared quite by accident. We had mated males raised in enriched environments with enriched females, and impoverished females with impoverished males, in order to examine the brains of the offspring (Diamond et al., 1971). The cortex of the enriched male parents was found to be 7% thicker than that of the impoverished male parents, as we had found in other experiments. However, upon comparing the cortical thickness of the enriched postpartum females with the impoverished postpartum females, we unexpectedly found *no* significant differences in an initial and a replication experiment. It became apparent upon close examination that the cortices in both the enriched and the impoverished postpartum females had increased in thickness during pregnancy, thus accounting for the nonsignificant differences between the two groups. In essence, the impoverished cortex had caught up with the enriched. These data led us to suspect that ovarian hormones might be affecting cortical structure.

A series of experiments followed, designed to clarify the role of ovarian hormones in cortical development and aging (Hoover and Diamond, 1976; Diamond et al., 1979; Pappas et al., 1979; Medosch and Diamond, 1982). At all times the data from both hemispheres were pooled, so that it became necessary to examine some of these data separately.

We compared the left and right cortical thickness differences in intact and ovariectomized females at 90 days of age (Diamond et al., 1981). The ovariectomized females were operated upon under cryogenic anesthesia at day 1, returned to their mothers until weaning at 21 to 23 days of age, then housed 3 to a cage until autopsy at 90 days of age. The intact females lived in similar conditions. As shown earlier (Table 9.1), at 90 days of age intact Long-Evans female rats possessed a *left* cortex thicker than the right in 7 of 9 areas, the exceptions being areas 2 and 18, but the differences were not statistically significant.

In the ovariectomized females at 90 days, we found that the *right* cortical mantle was thicker than the left in 7 out of 9 comparisons, with significant differences in areas 17, 18a, and 39. Specifically, the right hemisphere was greater in area 17 by 3% in 16 of 20 cases, $p < 0.02$; in area 18a, by 5% in 16 of 20 cases, $p < 0.001$; and in area 39, by 5% in 15 of 20 cases, $p < 0.02$. Table 9.3 indicates the direction of hemispheric differences.

We are now examining right and left cortices from animals treated with the contraceptive norethynodrel (primarily progesterone), estrogen, and progesterone (Hoover and Diamond, 1976; Pappas et al., 1979).

S_1 MALE RATS FROM DIFFERENT ENVIRONMENTAL CONDITIONS AND AGE GROUPS The S_1 strain of rats was developed by Robert Tryon at the University of California in the 1920s. Observing that some rats ran a maze better than others, he interbred the bright maze runners and the dull maze runners. The result was two new strains of rats, the maze-bright S_1 animals and the maze-dull S_3 animals. Male descendants of the former were used in these experiments.

At 90 days of age the standard colony S_1 males showed, in general, a right dominance in the cerebral cortical thickness, like the Long-Evans males (Diamond et al., 1981). However, the statistically significant differences were confined primarily to areas 17, 18a, and 39 in the S_1 strain, whereas in the Long-Evans strain significant right greater than left differences were seen throughout the cortex. These findings suggest that inbreeding for visual spatial cues had selectively strengthened the right dominance pattern only in some parts of the occipital cortex of the S_1 rats.

Let us examine individual cortical areas between the Long-Evans and S_1 strains. *At every age* in both strains the right area 17 is thicker than the left. However, in area 18, adjacent to 17, in the S_1 strain the left side is thicker than the right in over 50% of cases, whereas in the Long-Evans strain *at every age* the right area 18 is significantly

Table 9.3 Differences in right and left cortical thickness between intact and ovariectomized female Long-Evans rats at 90 days of age.

Subjects	N	Cortical area								
		10M	10L	4	3	2	18	17	18a	39
Intact animals	19–20									
Laterality		L>R	L>R	L>R	L>R	R>L	R=L	L>R	L>R	L>R
Significance		N.S.	N.S.	N.S.	N.S.	N.S.	=	N.S.	N.S.	N.S.
Ovariectomized animals	18									
Laterality		R>L	R>L	R>L	L>R	R>L	L>R	R>L	R>L	R>L
Significance		N.S.	N.S.	N.S.	N.S.	N.S.	N.S.	0.02	0.001	0.02

thicker than the left. It is area 18 that is most susceptible to change when the S_1 rats are exposed to enriched environments (Diamond, 1976), at a time before the rest of the cortex shows any significant changes. We have not yet placed Long-Evans rats in different environments to determine whether their area 18 also responds before the rest of the cortex.

These data show that patterns of cortical asymmetry in adjacent cortical areas can proceed in opposite directions in different strains. Not only do some areas show laterality patterns which vary between strains, but exposing rats to different environments can affect adjacent cortical regions differently. The environmental conditions we use in the laboratory have been published previously (Diamond et al., 1972; Diamond, 1976) and will not be given in detail here. They include an enriched condition (EC), the standard colony condition (SC), and an impoverished condition (IC).

Figure 9.1A illustrates the cortical thickness differences in area 17 of S_1 rats exposed to EC, SC, and IC for different periods of time at different ages: 60 to 64 days, 60 to 90 days, 25 to 55 days, and 25 to 105 days. At every age and in every condition the right area 17 is thicker than the left. With enrichment, thickness increases in both the right and left hemispheres; with impoverishment, the cortex in both hemispheres decreases in thickness.

Let us now look at area 18 in Fig. 9.1B. In every age group the left-sided area is larger than the right with enrichment; both sides are thicker than those of the standard colony animals. Cortical thickness in the impoverished animals is less than that of the standard colony, but the pattern of change is not so consistent as that of the enriched animals.

Contrast now area 2 (Fig. 9.1C), which in the developmental and aging study of the Long-Evans rats did not develop significant right

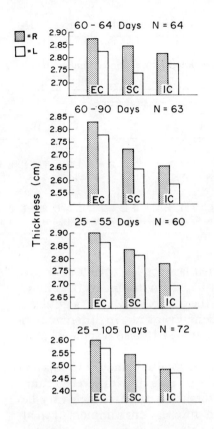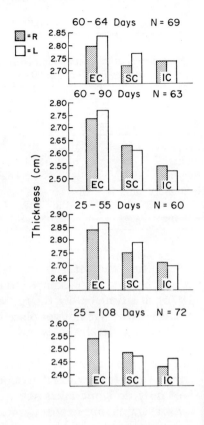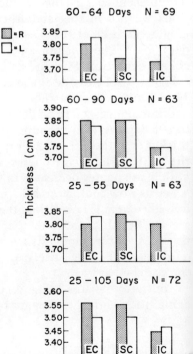

Fig. 9.1 Cortical thickness on right and left sides in different cortical areas in Long-Evans male rats raised under different environmental conditions. A, area 17; B, area 18; C, area 2. EC = enriched environment; SC = standard colony environment; IC = improved environment.

greater than left changes until later in life. In the S_1 strain we see an extremely varied pattern—not only in the SC, but in the EC and IC as well at different ages. There is no predictable pattern here at all.

Right-Left Hippocampal Differences

DEVELOPMENT AND AGING IN MALE AND FEMALE RATS Having studied cortical asymmetry in Long-Evans rats, we wondered whether such well-defined sex differences extended to subcortical structures. We first examined right-left thickness differences in the dorsal hippocampus of 90-day-old males and females (Diamond et al., 1982). The right hippocampus was indeed thicker than the left in males, but difference was not statistically significant. On the other hand, the left hippocampus was significantly thicker than the right in the 90-day-old female ($p < 0.003$).

We then carried out a more extensive analysis of the dorsal hippocampus in Long-Evans male rats of the following ages: 6, 10, 14, 26, 41, 77, 90, 108, 185, 300, 400, and 650 days (Diamond et al., 1982). The right male hippocampus was thicker than the left in 13 of 14 age groups (with nonsignificant differences at 77, 90, and 300 days); the reversal at 400 days of age was nonsignificant.

We were surprised to find that patterns of structural asymmetry in the male cerebral cortex and hippocampus change specifically around the period of sexual maturity. The right greater than left pattern diminished in the hippocampus between 41 and 108 days of age; in the cortex it increased between 55 and 90 days of age. Since we know from our studies of environmental influences on the cortex (Diamond, 1976) that measurable changes in thickness can take place in only 4 days, changes in brain structure could easily occur during early sexual maturity. But why left-right changes should take place remains a mystery.

We could find no mention elsewhere in the literature of changes in hippocampal asymmetry with age (see for example Loy and Milner, 1980, and Stokes and McIntyre, 1981). We have previously speculated (Diamond et al., 1982) on whether the role of the hippocampus in forming an environmental map, as reported by others, is relevant to our results. The map-building function might be related to the lateral visual cortex, because the right-left patterns during development and aging follow very similar directions in this cortex and in the hippocampus. Interesting connections between these two areas have been reported (Jones and Powell, 1970).

The female dorsal hippocampus was measured in 6 age groups: 7,

14, 21, 90, 180, and 390 days (Diamond et al., 1983). At every age but one (7 days), the left side was thicker than the right, but the differences were statistically significant only at 21 and 90 days. The patterns of hemispheric change for males and females are shown side by side in Fig. 9.2 for comparison. In the female there is a decreasing left-right difference in hippocampal thickness with aging, except at 90 days when the difference is both significant and striking. In the male, on the other hand, the right-left differences in hippocampal thickness are very marked in the early part of the animal's life, but they also decrease with aging. In both sexes the hippocampal right-left difference is almost nonexistent by the time the animals reach 390 to 400 days of age. A systematic study of the right and left hippocampus at different ages is essential for understanding the advantages of laterality or dominance.

Let us summarize the topics presented.

(1) The right and left cerebral cortical thickness patterns during development and aging in Long-Evans males and females are different. In males the right cortex is, on the whole, thicker than the left; in females the opposite is true, but the individual differences do not reach statistical significance. In the male, laterality in the cortex decreases in very old age.

(2) The left-right cerebral cortical thickness patterns differ between intact and gonadectomized male Long-Evans rats. Occipital cortex does not change its right greater than left pattern after neonatal gonadectomy, but more rostral cortex does.

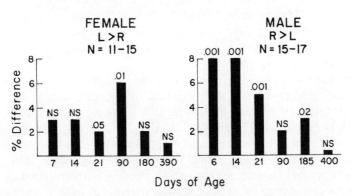

Fig. 9.2 Percent differences in dorsal hippocampal thickness on right and left sides in male and female Long-Evans rats at different ages. N = range of values of numbers of animals in different studies at each age; NS or numbers above each column indicate significance level of right-left difference.

(3) The left-right cerebral cortical thickness patterns differ between intact and ovariectomized female Long-Evans rats. After neonatal ovariectomy the right occipital cortex becomes significantly thicker than the left by 90 days of age, and the more rostral regions also show a right greater than left pattern in contrast to that of intact female littermates of similar age.

(4) The right and left cerebral cortical thickness patterns in S_1 males from 3 separate environments and 4 different age groups are varied. In area 17, the right greater than left pattern is consistent regardless of environmental condition. The left area 18 is generally larger than the right in animals exposed to the enriched environment. In area 2, the left-right pattern varies with age and environment; in other words, no consistency is seen.

(5) The right and left hippocampal thickness patterns during development and aging in Long-Evans males and females are quite different from those of the cortex. The male has large, significant differences in early life, which decrease considerably with aging. In the female, the left hippocampus is thicker than the right but the differences reach statistical significance only at 21 and 90 days of age. The differences also decrease by 390 days of age.

We conclude that laterality of the cerebral cortex and hippocampus in Long-Evans and S_1 rats varies with species, sex, age, and environment.

References

Diamond, M. C. 1976. Anatomical brain changes induced by environment. In L. Petrinovich and J. McGaugh, eds., *Knowing, Thinking and Believing*. New York: Plenum Press, pp. 215-241.

Diamond, M. C., Johnson, R. E., and Ingham, C. 1971. Brain plasticity induced by environment and pregnancy. *Internat. J. Neurosci.* 2: 171-178.

Diamond, M. C., Rosenzweig, M. R., Bennett, E. L., Lindner, B., and Lyon, L. 1972. Effects of environmental enrichment and impoverishment on rat cerebral cortex. *J. Neurobiol.* 3:47-64.

Diamond, M. C., Johnson, R. E., and Ingham, C. A. 1975. Morphological changes in young, adult, and aging rat cerebral cortex, hippocampus and diencephalon. *Behav. Biol.* 14:163-174.

Diamond, M. C., Johnson, R. E., and Ehlert, J. 1979. A comparison of cortical thickness in male and female rats — normal, gonadectomized, young and adult. *Behav. Neurol. Biol.* 26:485-491.

Diamond, M. C., Dowling, G. A., and Johnson, R. E. 1981. Morphologic cerebral cortical asymmetry in male and female rats. *Exp. Neurol.* 71:261-268.

Diamond, M. C., Murphy, G. M., Jr., Akiyama, K., and Johnson, R. E. 1982. Morphologic hippocampal asymmetry in male and female rats. *Exp. Neurol.* 76:553–566.

Diamond, M. C., Johnson, R. E., Young, D., and Singh, S. S. 1983. Age related morphologic differences in the rat cerebral cortex and hippocampus: male-female; right-left. *Exp. Neurol.* 81:1–13.

Dowling, G. A., Diamond, M. C., Murphy, G. M., Jr., and Johnson, R. E. 1982. A morphologic study of male rat cerebral cortical asymmetry. *Exp. Neurol.* 75:51–67.

Hoover, D. M., and Diamond, M. C. 1976. The effects of norethynodrel administration on the rat visual cortex exposed to differential environment: a preliminary study of electrolytes and water. *Brain Res.* 103:139–142.

Jones, E. G., and Powell, T. P. S. 1970. An anatomical study of converging sensory pathways within the cerebral cortex of the monkey. *Brain* 93:793–820.

Krieg, W. J. S. 1946. Connections of the cerebral cortex. I. The albino rat. A. Topography of the cortical areas. *J. Comp. Neurol.* 84:221–275.

Loy R., and Milner, T. 1980. Sexual dimorphism in extent of axonal sprouting in rat hippocampus. *Science* 208:1282–83.

Medosch, M. M., and Diamond, M. C. 1982. Rat occipital cortical synapses after ovariectomy. *Exp. Neurol.* 75:120–133.

Pappas, G. T. E., Diamond, M. C., and Johnson, R. E. 1979. Morphological changes in the cerebral cortex of rats with altered levels of ovarian hormones. *Behav. Neurol. Biol.* 26:298–310.

Sherman, G. F., Garbanati, J. A., Rosen, G. D., Yutzey, D. A., and Denenberg, V. H. 1980. Brain and behavior asymmetries for spatial preference in rats. *Brain Res.* 192:61–67.

Stokes, K. A., and McIntyre, D. C. 1981. Lateralized asymmetrical state-dependent learning produced by kindled convulsions from rat hippocampus. *Physiol. Behav.* 26:163–169.

Webster, W. G. 1977. Territoriality and the evolution of brain asymmetry. *Ann. N.Y. Acad. Sci.* 299:213–221.

Chapter 10

Functional and Neurochemical Asymmetries

Stanley D. Glick
Raymond M. Shapiro

Surprising to many people—in fact, initially greeted with some disbelief—has been the recent evidence that rats, like humans, have lateralized brains. Historically, studies of brain lateralization in humans have focused on hemispheric differences in language functions and on handedness. It is not difficult to understand why the possibility of a lateralized rat brain has aroused skepticism.

Nigrostriatal Asymmetry and Circling

It was the observation of a motor (or perhaps more accurately, a sensorimotor) asymmetry, the relationship of which to handedness is still not clear, that led to the initial characterization of cerebral asymmetry in rats. Normal rats were shown to turn in circles, either spontaneously at night (Glick and Cox, 1978) or after drugs during the day (Jerussi and Glick, 1976), in a preferential direction, some rats being consistently right-sided and others consistently left-sided (Glick et al., 1977a). At the time this observation was first made (Jerussi and Glick, 1974), it was well known that unilateral damage of the dopamine-containing nigrostriatal system (that is, the substantia nigra, nigrostriatal bundle, and corpus striatum) would also cause rats to circle toward the lesion, and that such circling or rotation could be potentiated by dopaminergic drugs (Ungerstedt, 1971; Glick et al., 1976; Pycock, 1980). As the lesion-induced effects were attributable to an imbalance in striatal dopaminergic function, it was postulated that turning in normal rats, though of considerably less intensity than in lesioned rats, would similarly be attributable to a striatal dopamine imbalance. Indeed, it was eventually demonstrated that the concentrations of dopamine in the two striata normally differed by about 15% (Zimmerberg et al., 1974) and that high doses of d-amphetamine (20 mg/kg, intraperitoneal) increased this asymmetry to approximately 25% (Glick et al., 1974) while inducing rats to rotate contralateral to the side with the higher dopamine levels.

The endogenous asymmetry in nigrostriatal function appeared to be related to spatial behavior in general. The normal left or right preferences of rats in a two-lever operant chamber (Glick and Jerussi, 1974) or in a T-maze (Zimmerberg et al., 1974) were correlated in direction with d-amphetamine-induced rotation. Neurochemical experiments demonstrated that the dopamine level was significantly higher in the striatum contralateral to the rat's side preference (Zimmerberg et al., 1974). Electrical stimulation of the striatum ipsilateral to the side preference resulted in the rat's switching sides for the duration of stimulation (Zimmerberg and Glick, 1975). All of these results suggest that rotation is a stereotyped form of spatial behavior and that spatial biases derive, at least in part, from a nigrostriatal asymmetry.

The fact that normal rats rotate and have side preferences means that the degree of rotation following a unilateral lesion should vary depending upon whether the lesion is ipsilateral or contralateral to the preoperative direction of rotation. Indeed, for both apomorphine (Jerussi and Glick, 1975) and d-amphetamine (Glick, 1976) it was shown that the intensity of rotation (full turns per hour) was twice as large in rats with striatal lesions ipsilateral to the preoperative direction as in rats with contralateral lesions. Differences in the effects of striatal lesions on the two sides have also been demonstrated in other contexts. Striatal lesions ipsilateral to side preferences facilitated rats' timing performance, whereas contralateral lesions impaired it (Glick and Cox, 1976). Lesions ipsilateral and contralateral to side preferences facilitated and impaired, respectively, passive avoidance learning (Rothman and Glick, 1976). The side of a lesion is obviously an important variable to be considered in assessing its effect. Many of the conflicting findings generated by studies utilizing unilateral lesions (Glick et al., 1976; Pycock, 1980) can probably be attributed to the failure to account for endogenous directional biases.

Other Relationships of Sidedness

Although the study of rodents turning in circles might seem to have little potential relevance for clinical medicine, several considerations taken together suggest that sidedness intimately modulates normal behavior. Comparisons of rats having strong or weak directional biases showed that the strength of normal biases was related generally to overall learning ability (Glick et al., 1977b), as well as specifically to the ability to discriminate left from right (Zimmerberg et al., 1978). Rats lacking clear spatial biases were found to be hyperactive, to have difficulty learning a variety of tasks, and to be unable to distinguish left from right. It was proposed (Glick et al., 1977a) that nonlateralized or weakly lateralized rats may represent a model of minimal brain dysfunction (attention deficit disorder, or hyperkinetic syndrome). Affected children are characterized by hyperactivity, learning difficulties and, frequently, poor cerebral dominance (Gazzaniga, 1973); motor disabilities as well as electrophysiological abnormalities may occur asymmetrically (Reitan and Boll, 1973; Conners, 1973.) Amphetamines and related drugs have been found to be effective treatments, especially when there is evidence of brain damage (Satterfield et al., 1973; Millichap, 1973; Shekim et al., 1979). It is possible that these therapeutic effects are attributable to enhancement of nigrostriatal asymmetry and/or cerebral dominance generally.

Iadarola and Gale (1982) have proposed that nigrostriatal dysfunction may underlie some seizure disorders. The Mongolian gerbil, an animal susceptible to handling-induced seizures, has been used in our laboratory to demonstrate that rotational behavior and seizure proneness are inversely related (Schonfeld and Glick, 1981). Pharmacological (dopamine agonists, prototypical antiepileptics) or surgical (unilateral striatal lesions) treatments that increased rotation were found to decrease seizure activity, whereas a treatment (haloperidol) that decreased rotation was observed to exacerbate seizure activity. These data suggest than an understanding of some seizure disorders, as well as the drugs used to treat them (Frumkin and Grim, 1981), may account for functional changes in the two hemispheres separately.

In an entirely different area, some recent data indicate a relationship between sidedness and fertility in rats. Litter size was found to vary with the left or right bias of the parents. Left-biased females had significantly larger (t-test, $p < 0.05$) litters than right-biased females (mean ± S.D. pups per litter: 9.46 ± 3.59, $N = 40$ versus 7.75 ± 3.58, $N = 34$); in contrast, litters were significantly larger (t-test, $p < 0.05$) when the male parent was right-biased than when left-biased (mean ± S.D. pups per litter: 9.55 ± 3.63, $N = 36$ versus 7.85 ± 3.55, $N = 38$). Thus the pairing of a left-sided female with a right-sided male resulted in the largest litters (10.00 ± 3.46,

N = 23); the pairing of a right-sided female with a left-sided male resulted in the smallest litters (7.19 ± 3.34, N = 21); and the pairings of same-sided parents resulted in litters of intermediate size (left-left parents: 8.71 ± 3.72, N = 17; right-right parents: 8.69 ± 3.90, N = 13). These results suggest an association between cerebral asymmetry and regulation of the gonads. Gerendai and Halasz (1981) reviewed evidence that the two sides of the hypothalamus are differentially affected by unilateral alterations of gonadal function. Nordeen and Yahr (1982) reported that the two sides of the neonatal hypothalamus are differentially responsive to unilateral estrogen pellet implants. In view of other evidence relating a hypothalamic asymmetry to the nigrostriatal asymmetry (Glick et al., 1979, 1980c), the present data suggest that some of the interindividual variation in normal neuroendocrine function may be determined by asymmetric relationships between corresponding hypothalamic nuclei on opposite sides of the brain.

Population Bias and Cortical Asymmetry

Until recently we believed that left-right biases in rotation and side preference occurred randomly in a rat population. In a typical experiment, for example, 50% of rats rotated to the right and 50% to the left. But in sharp contrast were several findings of left-right asymmetry in the rat cortex: differences in cortical thickness (for example, Diamond et al., 1975), differences in behavioral effects of left and right cortical lesions (Denenberg et al., 1978; Robinson, 1979), and a difference in frontal cortical energy metabolism (Glick et al., 1979). The results of the latter study suggested that the left-right frontal cortex asymmetry modulated the nigrostriatal asymmetry: side preferences were greater if frontal cortical deoxyglucose uptake was higher on the contralateral side. Because the left frontal cortex was usually more active than the right, it was reasoned that in a large population more rats should have right-side preferences than left-side preferences, and right preferences should be greater than left preferences. We therefore reviewed data on 602 rats (all female) tested for nocturnal or d-amphetamine-induced (1.0 mg/kg) rotation and found a small (54.8%) but significant ($p < 0.025$) right population bias. Furthermore, right-sided rats were more active and had stronger side preferences than left-sided rats (Glick and Ross, 1981b). If the frontal cortex modulates spatial bias, then bilateral lesions of the frontal cortex, by removing this modulation, should decrease side preferences and activity in right-sided rats and increase these parameters in left-sided rats. This prediction was also verified (Ross and Glick, 1981).

More recently, we have observed sex-dependent differences between left-sided and right-sided rats in cocaine-induced (20.0 mg/kg) rotation (Glick et al., in preparation): right-biased (N = 14) females rotated more than left-biased (N = 13) females (mean net rotations ± S.E. per hour: 179.4 ± 18.1 versus 89.1 ± 15.8, t-test $p < 0.01$), whereas left-biased (N = 19) males rotated more than right-biased (N = 13) males (126.6 ± 18.9 versus 43.7 ± 11.1, t-test, $p < 0.01$). Overall, female rats were more sensitive to cocaine than male rats. Although similar sex differences are known to occur with d-amphetamine (Brass and Glick, 1981; Becker et al., 1982), differences between left-biased and right-biased rats were comparatively small. The direction of rotation elicited by either drug in an individual rat was the same, but the quantitative effects of the two drugs were not correlated with each other.

Cocaine is a potent inhibitor of dopamine reuptake in the striatum (Snyder and Coyle, 1969; Heikkila et al., 1975). This action has been shown to be responsible for the efficacy of cocaine in inducing rotation in rats with unilateral lesions of substantia nigra (Heikkila et al., 1979a,b). It appears likely that the same mechanism is involved in the rotation of naive rats, although cocaine also inhibits reuptake of norepinephrine and serotonin (Ross and Renyi, 1967a,b) and it is possible that these actions also contribute (Glick et al., 1976; Pycock, 1980). The maximum effect of cocaine was at least 30% greater than ever observed with other drugs. This extreme efficacy, along with the rather remarkable differences between left-sided and right-sided rats, suggests that cocaine is affecting a fundamental mechanism involved in the regulation of brain asymmetry. This mechanism may have to do specifically with the importance of dopamine reuptake for the regulation of dopamine neuron activity and hence striatal dopaminergic asymmetry, or more generally with the importance of monoamine uptake in multiple brain regions and neurotransmitter pathways. Sex differences in anatomical asymmetries have been reported in the cortex (Diamond et al., 1981) and the hippocampus (Diamond et al., 1982). In both structures the left sides are usually thicker in female rats and the right sides are usually thicker in males. As both the cortex (Ross and Glick, 1981) and the hippocampus (Glick et al., 1980a) have been shown to modulate the intensity of rotation, it is not improbable that the uniqueness of cocaine's effects on rotation is attributable to actions in structures other than the striatum.

Heritable and Prenatal Determinants

In contrast to the right-sided population bias in our Sprague-Dawley rats, other investigators have observed left biases in Purdue-Wistar rats (Sherman et al., 1980) and in ICR mice (Korczyn and Eshel, 1979). These differences suggest a role for genetic factors in determining the direction of an individual rat's side preference. A subsequent study provided more direct evidence for such a heritable influence (Glick, 1983).

Adult male and female rats (Sprague-Dawley), identified as either left or right rotators, were mated in such a way that all possible combinations of phenotypes would be represented: left female with right male, left female with left male, and so on. Pups were weaned at 21 days after birth, the sexes separated, and the males and females of each litter housed in standard group cages until testing for nocturnal rotation as adults (Glick and Cox, 1978). Data were obtained from 70 litters having a total of 235 female and 220 male offspring. Overall, the side preferences of male offspring were significantly (chi-square tests) like those of the male parent and unlike those of the female parent: 60.0% ($p < 0.005$) of the males rotated in the same direction as the male parent and 37.3% ($p < 0.0005$) in the same direction as the female parent. In contrast, there were no significant tendencies for the female offspring: 50.6% rotated in the same direction as the male parent and 47.7% in the same direction as the female parent.

Separate analysis of the results of each mating type confirmed that parental bias was clearly related (chi-square test, $p < 0.0005$) to the left-right distribution of male offspring. When the parents rotated in opposite directions, approximately 70% of the male offspring rotated in the same direction as the male parent and opposite the direction of the female parent. When both parents rotated in the same direction, there was no tendency for the male offspring to resemble or not resemble the parents. It appeared that the male and female influences contributed equally but oppositely to the bias of the male offspring and effectively canceled each other when the parents had same-sided biases. Female offspring were distributed randomly in all instances.

It was not immediately apparent why male and female parents had opposing effects on the male offspring or why there were no heritable effects on the female offspring. One factor might be in utero exposure to testosterone. Prenatal and early postnatal exposure of rodents to androgens is known to masculinize the central nervous system (MacLusky and Naftolin, 1981). A major source of androgen in fetal females is thought to be testosterone derived from male littermates (Clemens, 1974; vom Saal and Bronson, 1978; Meisel and Ward, 1981). How testosterone interacts with the developing nervous system to produce a permanent differentiating effect is largely unknown. One hypothesis is that testosterone induces a change in gene expression (MacLusky and Naftolin, 1981). In the present context it was postulated that testosterone reverses the coding, in terms of cerebral laterality, of the heritable female influence. In the absence of fetal exposure to testosterone, it might be predicted that a tendency of the offspring to resemble the female parent would be detected; in the presence of large amounts of testosterone, the reverse would be predicted. Since male fetuses produce testosterone, the only female influence detectable in male offspring would be the tendency to have biases opposite the direction of the female parents.

The amounts of testosterone in the blood and amniotic fluid of female fetuses (vom Saal and Bronson, 1980), as well as the degree of female masculinization in adulthood (Clemens, 1974; Clemens et al., 1978; vom Saal and Bronson, 1978, 1980; Meisel and Ward, 1981), are related to intrauterine proximity to male fetuses. The higher the proportion of males in a litter, the greater the likelihood that the litter contains females situated in close proximity to males in utero. The tendencies of the female offspring to resemble or not resemble the female parent should be related, therefore, to a lower or higher percentage of males in the litter. Analysis of the data support this reasoning. The distribution of female offspring with respect to the female parent varied with the relative sex frequencies of the litter. In litters having more males than females, there was a significant tendency (chi-square test, $p < 0.02$) for the female offspring to rotate in the opposite direction from the female parent, whereas there was a slight opposite trend in litters having more females than males. Overall, the percentage of females per litter that rotated in the same direction as the female parent was inversely correlated with the number of males in the litter ($r = -0.28$, $p < 0.05$) and with the percentage of males per total litter offspring ($r = -0.33, p < 0.01$).

The differences in how the male and female influences are transmitted suggest involvement of the sex chromosomes. However, an interaction between gene expression and hormonal modulation would make it difficult to discern any simple pattern of inheritance. Moreover, in addition to genes and hormones environmental pressures also affect the direction and/or degree of sidedness (Collins, 1975; Denenberg, 1981); there may be a critical period of optimal plasticity (McGlone, 1980; Glick et al., 1980b). The origins of sidedness in rodents, and possibly of handedness in humans, undoubtedly involve the complex interplay of multiple determinants. The present data indicate only that one of these determinants is heritable in nature.

Multiple Asymmetries and Developmental Changes

The labeled 2-deoxy-D-glucose (dGlc) technique (Sokoloff et al., 1977) can be used to study the functional state of the nervous system in a variety of experimental conditions. Based on the expectation that cerebral asymmetry might be manifested as differences in metabolic activity, we used a modified dGlc technique (Meibach et al., 1980) to evaluate potential asymmetries in several brain regions (Glick et al., 1979). The results indicated that hemispheric asymmetry in the rat is present in several brain regions. Three different kinds of asymmetry were identified: (1) differences in activity (dGlc uptake) between sides of the brain contralateral and ipsilateral to the direction of rotation (midbrain and striatum); (2) differences in activity between left and right sides (frontal cortex and hippocampus); and (3) absolute differences in activity between sides that were correlated with the rate of rotation (thalamus and hypothalamus) or with random movement (cerebellum). Cerebral asymmetry in the rat is not, therefore, simply a case of one side of the brain being consistently "dominant" over the other. Different asymmetries are organized along different dimensions in different structures. This is probably true in humans as well (Glick et al., 1982a).

The dGlc technique was also used to investigate the ontogeny of cerebral laterality in rats (Ross et al., 1981, 1982). Newborn rats had left-right asymmetries in dGlc uptake in many structures and these asymmetries changed during development. In females there were significant left-to-right gradients in brainstem and midbrain and right-to-left gradients in hippocampus and diencephalon; for example, the hippocampus was right biased in the neonate and left biased in the adult. The only gradient observed in males was a right-to-left gradient in midbrain. Perhaps the most intriguing finding was a significant relationship between brain activity (dGlc uptake) and brain asymmetry: the more active a structure relative to the rest of the brain, the more likely that the structure was right biased and vice versa. This relationship occurred in most brain regions and in both sexes, although there were significant differences between sexes in the slopes of the relationships in some structures (cortex, striatum). The finding implies that left-right asymmetry is intimately and dynamically related to changes in functional activity of a structure; treatments that preferentially increase or decrease activity of a structure should have predictable effects on the left-right asymmetry of that structure.

In both sexes we also observed neonatal asymmetries in tail position (Ross et al., 1981) that were predictive of adult turning preferences. A possibly related finding in humans was reported by Michel (1981): an asymmetry in neonatal head posture predicted handedness at 16 and 22 weeks of age. As with handedness in humans (McGlone, 1980), there are significant differences between the sexes in the strength of side preferences in rats (Glick et al., 1980b; Robinson et al., 1980; Brass and Glick, 1981; Becker et al., 1982).

Neuroanatomical (Galaburda et al., 1978) and functional (Witelson, 1976) findings in humans suggest that cerebral lateralization is present at birth, changes during development, and is sexually dimorphic. These same characteristics appear to be exhibited by rats as well. Corballis and Morgan (1978) hypothesize that the development of cerebral lateralization in the human and other species may be attributable to a left-right maturational gradient: the left hemisphere presumably develops earlier and/or more rapidly than the right. The above dGlc data indicate that maturational gradients do indeed exist, but that such gradients are structure specific and sexually dimorphic.

Reward Mechanisms in Drug-Induced Euphoria

Neurological findings have suggested that the two sides of the human brain are specialized for affect, one hemispheric characterized as more joyful and the other as more depressive (Gainotti, 1972; Galin, 1974; Sackeim et al., 1982). Based on the premise that differences in affect result from differences in the activity of the mechanisms mediating reinforcement, we postulated and succeeded in demonstrating that the two sides of the rat brain are differentially sensitive to reinforcing lateral hypothalamic stimulation (Glick et al., 1980c). Bipolar stainless steel electrodes were implanted in both lateral hypothalami of naive female Sprague-Dawley rats. The rats were subsequently placed in operant chambers and allowed to self-stimulate (0.5 sec, 60 Hz sine wave, 10–150 μA), during which time rotation was also measured. All rats had asymmetries in self-stimulation thresholds related to the preferred direction of rotation: thresholds were lower on the contralateral side, and overall rate-intensity functions were generally higher and displaced leftward on that side.

The finding of a self-stimulation asymmetry suggested that differences in the reinforcing qualities of different drugs might be due to their differential effects on reward mechanisms in the two sides of the brain. This idea was initially pursued by examining the effects of d-amphetamine and morphine in the self-stimulation paradigm. Various doses of each drug were administered twice, once with the left side of the brain stimulated first and once with the right side stimulated first. The effects of the two drugs with respect to lateralization were quite different. Whereas d-amphetamine preferentially affected the low-threshold side of the brain, low doses of morphine (2.5 mg/kg, for instance) preferentially affected the high-threshold side. The low-dose effects of both drugs were to shift the rate-intensity functions to the left; with d-amphetamine, however, this shift occurred more for the side contralateral to the direction of rotation, while with morphine it occurred more for the ipsilateral side. The dose response curve for d-amphetamine was an inverted U function (Glick et al., 1981), with high doses (such as 2.0 mg/kg) producing less of an increase in reward sensitivity and asymmetry than lower doses (such as 0.25 to 0.5 mg/kg). On the other hand, high doses (10.0 to 20.0 mg/kg) of morphine augmented reinforcement thresholds and shifted rate-intensity functions rightward; however, this depressant effect was more pronounced in the low-threshold side of the brain so that the morphine-induced reversal of reward asymmetry was fairly constant across doses (Glick et al., 1982b). The facilitatory and depressant actions of morphine on lateral hypothalamic self-stimulation preferentially occurred, therefore, on opposite sides of the brain.

We have begun investigating the possibility of self-stimulation asymmetries in other brain regions. Some interesting results have been observed with electrode placements in the dorsal hippocampus. An asymmetry in self-stimulation sensitivity was evident in the dorsal hippocampus, but in contrast to the lateral hypothalamus, the hippocampal asymmetry was unrelated to the direction of rotation. The hippocampal asymmetry may be organized with respect to left versus right: the left hippocampus had a lower self-stimulation threshold than the right hippocampus in 9 of 11 rats tested. In our deoxyglucose studies (Glick et al., 1979), the left hippocampus was also found to be more active than the right.

In contrast to the robust effects of d-amphetamine and morphine on lateral hypothalamic self-stimulation, both drugs had negligible effects on hippocampal self-stimulation in either side of the brain. However, two hallucinogens, LSD and phencyclidine (PCP), while having little or no effects on lateral hypothalamic self-stimulation, had substantial effects on hippocampal self-stimulation. Optimal doses of the latter drugs enhanced rates and lowered thresholds, though preferentially affecting opposite sides of the brain: whereas LSD (0.25 mg/kg) had larger effects on the normally more sensitive side, PCP (2.5 mg/kg) had larger effects on the normally less sensitive side. LSD enhanced the hippocampal asymmetry, while PCP reduced or reversed it. Thus, the mechanisms of action of four commonly abused drugs can be dissociated in terms of both their site of action and side of action.

The finding that reward processes are lateralized may have some relevance to our understanding of psychopathological states. Numerous observations indicate, for example, that the dominant hemisphere is overactive in schizophrenia (Flor-Henry, 1976; Gur, 1978). Other data suggest that disturbances of reward mechanisms underlie schizophrenia (Stein and Wise, 1971; Wise et al., 1978), that amphetamine potentiates or induces motor and delusional symptoms of schizophrenia (Angrist and Gershon, 1970, 1972) and that dopaminergic systems in brain are hyperactive in schizophrenia (Snyder, 1976). Schizophrenia may to some extent be a disorder of a hyperactive dopaminergic pathway in the dominant hemisphere, and amphetamine may exacerbate or induce the disorder by selectively affecting the dominant side. "Dominance" may be attributable to a lower threshold for excitation and, as in the self-stimulation paradigm, amphetamine may enhance dominance by inducing a lateralized increase in reward sensitivity. Schizophrenic hallucinations may be mediated by abnormal activity in several brain regions, apparently including the globus pallidus (a probable site of amphetamine action), but more importantly the hippocampus and other limbic structures (Horowitz and Adams, 1970). The latter is intriguing in relation to our observations that LSD and PCP alter hippocampal asymmetry in reward sensitivity. These two drugs induce visual and auditory hallucinations, respectively (Martin, 1977), and the mechanisms of both types of hallucinations appear to be lateralized in the human brain (Penfield and Perot, 1963).

Striatal Dopamine Receptors

One component of mechanisms mediating lateralized drug actions in the brain may be an asymmetry in the distribution of neurotransmitter receptors. We have sought to determine if there are differences in dopamine receptors in the two sides of the rat brain. [³H]Apomorphine was employed as a radioligand in binding studies with striatum. This ligand was thought to be particularly relevant, since apomorphine itself elicits rotation (Jerussi and Glick, 1976). Preliminary experiments confirmed the findings of others (Sokoloff et al., 1980; Seeman, 1980) that [³H]apomorphine labels two distinct binding sites in rat striatum. The binding that is sensitive to 200 nM domperidone has been reported to be postsynaptic in rat striatum and has been classified as a D4 dopamine receptor. There were no differences between sides of the brain or between sexes in binding to this site.

However, there were differences in the [³H]apomorphine binding that occurs in the presence of 200 nM domperidone and is sensitive to 2 μM ADTN (2-amino-6,7-dihydroxy-1,2,3,4-tetrahydronaphthalene). This binding site, classified as a D3 dopamine receptor, has been reported to be largely presynaptic in rat striatum (Sokoloff et al., 1980; Seeman, 1980). There was significantly more total binding to the D3 site in the striata from females than in those from males (average difference 20.5%, analysis of variance, $p < 0.0003$). Other observations were that right rotators exhibited more total binding than left rotators, that the right side of the brain bound more ligand that the left side, and that there was more binding on the side of the brain ipsilateral to the directional bias of the animal (analysis of variance, $p < 0.05$ in each case).

The last three results were accounted for by the composite finding that right rotators showed a significant right/left asymmetry of binding, whereas the left rotators did not. That is, there was an equivalent amount of binding on both sides of the brain in left rotators, which was equivalent to the amount of binding on the left side of the brain in right rotators, but there was significantly greater binding (average 21.7%, paired t-tests, $p < 0.05$), in the right striata of right rotators of both sexes (Stollak and Glick, 1982). It remains to be determined whether the differences in [³H]apomorphine binding to the D3 site result from differences in the total number of binding sites and/or the affinity of the radioligand for the site.

The fact that significant differences between sexes and between sides of the brain were observed only with respect to the D3 site, even though the D3 and the D4 binding were assayed simultaneously, suggests that the mechanism(s) responsible for regulating asymmetry in striatal dopaminergic activity, as well as those responsible for sex differences in rotation (Glick et al., 1980b; Robinson et al., 1980; Brass and Glick, 1981; Becker et al., 1982), are predominantly presynaptic. This conclusion is consistent with other behavioral (Glick et al., 1977c) and biochemical (Jerussi et al., 1977) data.

It is now apparent that left-sided and right-sided rats differ behaviorally as well as neurochemically (Valdes et al., 1981). Behavioral and cerebral differences between left-handed and right-handed humans are well documented (for instance, by Rasmussen and Milner, 1977). Whereas right-handed humans almost always have left cerebral dominance, left-handed individuals are known to have either right cerebral, left cerebral, or ambivalent dominance (Hécaen and Sauguet, 1971). If underlying neurochemical asymmetries are related to cerebral dominance, neurochemical data from a random population of right-handers should exhibit clear left-right differences, but similar data from a random population of left-handers should exhibit no clear patterns of asymmetry. This certainly seems to be true for left-sided versus right-sided rats, at least with respect to the D3 striatal site. Though the similarity may only be superficial, we may have a model of the human situation that is worthy of further investigation.

Asymmetries in the Globus Pallidus

Our group was recently presented with an opportunity to determine whether neurochemical asymmetries were present in postmortem human brains. Rossor et al. (1980) measured indexes of 3 to 5 neurotransmitters in 9 structures, both left and right sides, of normal human brains. Because data on handedness or cerebral dominance were not available to these researchers, they did not attempt to assess the functional significance of any asymmetries. However, they kindly agreed to allow their data to be reanalyzed for exactly this purpose. Our basic premise was that if different neurochemical asymmetries are related to the same function, at least in terms of lateral bias, then there should be correlations among such asymmetries.

The reanalysis showed that human brain indeed has lateral asymmetries in several structures and transmitter systems (Glick et al., 1982a). Perhaps most impressive is the finding that left-right asymmetries in glutamic acid decarboxylase (GAD) and gamma-aminobutyric acid (GABA) were positively correlated in all 9 structures. GABA and GAD, indexes of the same transmitter system, were measured by different methods. Intriguing with respect to previous data are the findings that choline acetyltransferase (ChAT) and dopamine were both significantly left-biased in the globus pallidus; left-right asymmetries in ChAT and dopamine were also positively correlated with each other in the globus pallidus, as well as in the caudate nucleus and putamen. Since most of the patients (N = 14) should have been right-handed, pallidal ChAT and dopamine were obviously higher on the side contralateral to hand preferences. Similarly, in rats striatal dopamine levels were higher on the side contralateral to side preferences (Zimmerberg et al., 1974).

The "striatum" used for determining the dopamine asymmetry in rats (Zimmerberg et al., 1974) was dissected according to Glowinski and Iversen (1966) and therefore included the globus pallidus in addition to the caudate nucleus and putamen. Dissection of striatum for measuring deoxyglucose uptake (Glick et al., 1979) included only true striatum (caudate nucleus and putamen). The finding in human globus pallidus suggested that the difference in dissections might account for why, in rat striatum, the dopamine asymmetry (12% to 15%) was much greater than the deoxyglucose asymmetry (3% to 6%): as in humans, the largest asymmetry might simply be localized in the globus pallidus. Accordingly, $[1,2^3H]2$-deoxy-D-glucose uptake (Meibach et al., 1980) was selectively determined in rat caudate-putamen and globus pallidus. The side-to-side asymmetry in deoxyglucose uptake was 3.8% in caudate-putamen and 18.4% in globus pallidus (significantly different at $p < 0.001$, paired t-test), and only the latter asymmetry was significantly ($p < 0.05$, paired t-test) related to the direction of (N = 8) side preferences, the side with greater uptake being contralateral.

In recent anatomical experiments we measured the areas and estimated the volumes of caudate-putamen and globus pallidus of right-rotating and left-rotating rats (N = 13). Alternate 20 μm coronal sections of fresh frozen brain were stained with Sudan black and neutral red (Shapiro et al., in preparation) and then both sides of the caudate-putamen and globus pallidus were traced from enlarged projections. Whereas there was no significant asymmetry for the caudate-putamen, the left globus pallidus was significantly larger ($p < 0.001$, paired t-test) than the right; the mean left-right asymmetries for the two structures were 1.20% ± 0.61 S.E. and 6.52% ± 1.49 S.E., respectively. Further analysis revealed that preference during 24-hr rotation testing (Glick and Cox, 1978) was positively correlated with the left-right asymmetry for right-rotating rats and negatively correlated with the left-right asymmetry for left-rotating rats ($r = 0.82$ and -0.91, respectively; N = 7 and 6; $p < 0.05$ in both cases). Although these data must be regarded as preliminary because of the small number of subjects, the absence of any simple one-to-one relationship between the anatomy and the behavior should be noted. Numbers of neurons and/or relative sizes of homotypic structures are not necessarily related to functional activity (Agid et al., 1973). The data do, however, again draw attention to the differences between left-sided rats and right-sided rats and suggest that the globus pallidus may be fundamentally involved in such differences.

In this chapter we have reviewed experimental data indicating that the rat brain is asymmetric with respect to a variety of interrelated measures. It now appears that rat and human brains exhibit some similarities in the mechanisms of asymmetry, at least for certain structures. Moreover, there appears to be more reason than ever to suspect that studies in the rat will reveal functions of brain asymmetry that are relevant to man.

The research described in this chapter was supported by grants DA 01044 from NIDA and NS 14812 from NINCDS.

References

Agid, Y., Javoy, F., and Glowinski, J. 1973. Hyperactivity of remaining dopaminergic neurones after partial destruction of the nigrostriatal dopaminergic system in the rat. *Nature New Biol.* 245:150–151.

Angrist, B. M., and Gershon, S. 1970. The phenomenology of experimentally induced amphetamine psychosis: preliminary observations. *Biol. Psychiatry* 2:95–107.

Angrist, B. M., and Gershon, S. 1972. Psychiatric sequelae of amphetamine use. In R. I. Shader, ed., *Psychiatric Complications of Medical Drugs.* New York: Raven Press, pp. 175–199.

Becker, J. B., Robinson, T. E., and Lorenz, K. A. 1982. Sex differences and estrous cycle variations in amphetamine-elicited rotational behavior. *Europ. J. Pharmacol.* 80:65–72.

Brass, C. A., and Glick, S. D. 1981. Sex differences in drug-induced rotation in two strains of rats. *Brain Res.* 233:229–234.

Clemens, L. 1974. Neurohumoral control of male sexual behavior. In W. Montagna and W. A. Sadler, eds., *Reproductive Behavior.* New York: Plenum Press, pp. 23–53.

Clemens, L. G., Gladue, B. A., and Coniglio, L. P. 1978. Prenatal endogenous androgenic influences on masculine sexual behavior and genital morphology in male and female rats. *Horm. Res.* 10:40–53.

Collins, R. L. 1975. When left-handed mice live in right-handed worlds. *Science* 187:181–184.

Conners, K. 1973. Psychological assessment of children with minimal brain dysfunction. *Ann N.Y. Acad. Sci.* 205:283–302.

Corballis, M. C., and Morgan, M. J. 1978. On the biological basis of human laterality: I. Evidence for a maturational left-right gradient. *Behav. Brain Sci.* 1:261–336.
Denenberg, V. H. 1981. Hemispheric laterality in animals and the effects of early experience. *Behav. Brain Sci.* 4:1–49.
Denenberg, V. H., Garbanati, J., Sherman, G., Yutzey, D. A., and Kaplan, R. 1978. Infantile stimulation induces brain lateralization in rats. *Science* 201:1150–52.
Diamond, M. C., Johnson, R. E., and Ingham, C. A. 1975. Morphological changes in the young, adult and aging rat cerebral cortex, hippocampus and diencephalon. *Behav. Biol.* 14:163–174.
Diamond, M. C., Dowling, G. A., and Johnson, R. E. 1981. Morphologic cerebral cortical asymmetry in male and female rats. *Exp. Neurol.* 71:261–268.
Diamond, M. C., Murphy, G. M., Akiyama, K., and Johnson, R. E. 1982. Morphologic hippocampal asymmetry in male and female rats. *Exp. Neurol.* 76:553–565.
Flor-Henry, P. 1976. Lateralized temporal-limbic dysfunction and psychopathology. *Ann. N.Y. Acad. Sci.* 280:777–795.
Frumkin, L. B., and Grim, P. 1981. Is there pharmacological asymmetry in the human brain? An hypothesis for the differential hemispheric action of barbiturates. *Intern. J. Neurosci.* 13:187–197.
Gainotti, G. 1972. Emotional behavior and hemispheric side of the lesion. *Cortex* 8:41–55.
Galaburda, A. M., Lemay, M., Kemper, T. L., and Geschwind, N. 1978. Right-left asymmetries in the brain. *Science* 199:852–856.
Galin, D. 1974. Implications for psychiatry of left and right cerebral specialization. *Arch. Gen. Psychiatry* 31:572–583.
Gazzaniga, M. S. 1973. Brain theory and minimal brain dysfunction. *Ann. N.Y. Acad. Sci.* 205:89–92.
Gerendai, I., and Halasz, B. 1981. Participation of a pure neuronal mechanism in the control of gonadal functions. *Andrologia* 13:275–282.
Glick, S. D. 1976. Behavioral effects of amphetamine in brain damaged animals: problems in the search for sites of action. In E. Ellinwood, ed., *Cocaine and Other Stimulants.* New York: Plenum Press, pp. 77–96.
Glick, S. D. 1983. Heritable determinants of left-right bias in the rat. *Life Sci.*: in press.
Glick, S. D., and Cox, R. D. 1976. Differential effects of unilateral and bilateral caudate lesions on side preferences and timing behavior in rats. *J. Comp. Physiol. Psychol.* 90:528–535.
Glick, S. D., and Cox, R. D. 1978. Nocturnal rotation in normal rats: correlation with amphetamine-induced rotation and effects of nigrostriatal lesions. *Brain Res.* 150:149–161.
Glick, S. D., and Jerussi, T. P. 1974. Spatial and paw preferences in rats: their relationship to rate-dependent effects of D-amphetamine. *J. Pharmacol. Exp. Ther.* 188:714–725.

Glick, S. D., and Ross, D. A. 1981a. Lateralization of functions in the rat brain: basic mechanisms may be operative in humans. *Trends Neurosci.* 4:196–199.
Glick, S. D., and Ross, D. A. 1981b. Lateralized effects of bilateral frontal cortex lesions in rats. *Brain Res.* 210:379–382.
Glick, S. D., Jerussi, T. P., Waters, D. H., and Green, J. P. 1974. Amphetamine-induced changes in striatal dopamine and acetylcholine levels and relationship to rotation (circling behavior) in rats. *Biochem. Pharmacol.* 23:3223–25.
Glick, S. D., Jerussi, T. P., and Fleisher, L. N. 1976. Turning in circles: the neuropharmacology of rotation. *Life Sci.* 18:889–896.
Glick, S. D., Jerussi, T. P., and Zimmerberg, B. 1977a. Behavioral and neuropharmacological correlates of nigrostriatal asymmetry in rats. In S. Harnad, ed., *Lateralization in the Nervous System*. New York: Academic Press, pp. 213–249.
Glick, S. D., Zimmerberg, B., and Jerussi, T. P. 1977b. Adaptive significance of laterality in the rodent. *Ann. N.Y. Acad. Sci.* 299:180–185.
Glick, S. D., Jerussi, T. P., Cox, R. D., and Fleisher, L. N. 1977c. Pre- and post-synaptic actions of apomorphine: Differentiation by rotatory effects in normal rats. *Arch. Int. Pharmacodyn. Ther.* 225:303–307.
Glick, S. D., Meibach, R. C., Cox, R. D., and Maayani, S. 1979. Multiple and interrelated functional asymmetries in rat brain. *Life Sci.* 25:395–400.
Glick, S. D., Meibach, R. C., Cox, R. D., and Maayani, S. 1980a. Phencyclidine-induced rotation and hippocampal modulation of nigrostriatal asymmetry. *Brain Res.* 196:99–107.
Glick, S. D., Schonfeld, A. R., and Strumpf, A. J. 1980b. Sex differences in brain asymmetry of the rodent. *Behav. Brain Sci.* 3:236.
Glick, S. D., Weaver, L. M., and Meibach, R. C. 1980c. Lateralization of reward in rats: differences in reinforcing thresholds. *Science* 207:1093–95.
Glick, S. D., Weaver, L. M., and Meibach, R. C. 1981. Amphetamine enhancement of reward asymmetry. *Psychopharmacology* 73:323–327.
Glick, S. D., Ross, D. A., and Hough, L. B. 1982a. Lateral asymmetry of neurotransmitters in human brain. *Brain Res.* 234:53–63.
Glick, S. D., Weaver, L. M., and Meibach, R. C. 1982b. Asymmetrical effects of morphine and naloxone on reward mechanisms. *Psychopharmacology* 78:219–224.
Gur, R. E. 1978. Left hemisphere dysfunction and left hemisphere overactivation in schizophrenia. *J. Abnorm. Psychol.* 87:226–238.
Hécaen, H., and Sauguet, J. 1971. Cerebral dominance in left-handed subjects. *Cortex* 7:19–48.
Heikkila, R. E., Orlansky, H., and Cohen, G. 1975. Studies on the distinction between uptake inhibition and release of [^{3}H]dopamine in rat brain tissue slices. *Biochem. Pharmacol.* 24:847–852.
Heikkila, R. E., Cabat, F. S., and Duvoisin, R. C. 1979a. Motor activity and

rotational behavior after analogs of cocaine: correlation with dopamine uptake blockade. *Commun. Psychopharmacol.* 3:285–290.
Heikkila, R. E., Cabat, F. S., Manzino, L., and Duvoisin, R. C. 1979b. Rotational behavior induced by cocaine analogs in rats with unilateral 6-hydroxydopamine lesions of the substantia nigra: dependence upon dopamine uptake inhibition. *J. Pharmacol. Exp. Ther.* 211: 189–194.
Horowitz, M. J., and Adams, J. E. 1970. Hallucinations on brain stimulation: evidence for revision of the Penfield hypothesis. In W. Keup, ed., *Origin and Mechanisms of Hallucinations*. New York: Plenum Press, pp. 13–22.
Iadarola, M. J., and Gale, K. 1982. Substantia nigra: site of anticonvulsant activity mediated by γ-aminobutyric acid. *Science* 218:1237–40.
Jerussi, T. P., and Glick, S. D. 1974. Amphetamine-induced rotation in rats without lesions. *Neuropharmacology* 13:283–286.
Jerussi, T. P., and Glick, S. D. 1975. Apomorphine-induced rotation in normal rats and interaction with unilateral caudate lesions. *Psychopharmacologia* 40:329–334.
Jerussi, T. P., and Glick, S. D. 1976. Drug-induced rotation in rats without lesions: behavioral and neurochemical indices of a normal asymmetry in nigrostriatal function. *Psychopharmacology* 47:249–260.
Jerussi, T. P., Glick, S. D., and Johnson, C. L. 1977. Reciprocity of pre- and post-synaptic mechanisms involved in rotation as revealed by dopamine metabolism and adenylate cyclase stimulation. *Brain Res.* 129:385–388.
Korczyn, A. D., and Eshel, Y. 1979. Dopaminergic and non-dopaminergic circling activity of mice. *Neuroscience* 4:1085–88.
McGlone, J. 1980. Sex differences in human brain asymmetry: a critical survey. *Behav. Brain Sci.* 3:215–263.
MacLusky, N. J., and Naftolin, F. 1981. Sexual differentiation of the central nervous system. *Science* 211:1294–1303.
Martin, W. R. 1977. *Handbook of Experimental Pharmacology 45/II.* New York: Springer-Verlag.
Meibach, R. C., Glick, S. D., Ross, D. A., Cox, R. D., and Maayani, S. 1980. Intraperitoneal administration and other modifications of the 2-deoxy-D-glucose technique. *Brain Res.* 195:167–176.
Meisel, R. L., and Ward, I. L. 1981. Fetal female rats are masculinized by male littermates located caudally in the uterus. *Science* 213:239–242.
Michel, G. F. 1981. Right-handedness: a consequence of infant supine head-orientation preference. *Science* 212:685–687.
Millichap, G. L. 1973. Drugs in management of minimal brain dysfunction. *Ann. N.Y. Acad. Sci.* 205:321–334.
Nordeen, E. J., and Yahr, P. 1982. Hemispheric asymmetries in the behavioral and hormonal effects of sexually differentiating mammalian brain. *Science* 218:391–394.
Penfield, W., and Perot, P. 1963. The brain's record of auditory and visual experience—a final summary and discussion. *Brain* 86:596–696.

Pycock, C. J. 1980. Turning behaviour in animals. *Neuroscience* 5:461–514.
Rasmussen, T., and Milner, B. 1977. The role of early left-brain injury in determining lateralization of cerebral speech functions. *Ann. N.Y. Acad. Sci.* 299:353–369.
Reitan, R. M., and Boll, T. J. 1973. Neuropsychological correlates of minimal brain dysfunction. *Ann. N.Y. Acad. Sci.* 205:65–88.
Robinson, R. G. 1979. Differential behavioral and biochemical effects of right and left hemispheric cerebral infarction in the rat. *Science* 205:707–710.
Robinson, T. E., Becker, J. B., and Ramirez, V. D. 1980. Sex differences in amphetamine-elicited rotational behavior and the lateralization of striatal dopamine in rats. *Brain Res. Bull.* 5:539–545.
Ross, D. A., and Glick, S. D. 1981. Lateralized effects of bilateral frontal cortex lesions in rats. *Brain Res.* 210:379–382.
Ross, D. A., Glick, S. D., and Meibach, R. C. 1981. Sexually dimorphic brain and behavioral asymmetries in the neonatal rat. *Proc. Natl. Acad. Sci. USA* 78:1958–61.
Ross, D. A., Glick, S. D., and Meibach, R. C. 1982. Sexually dimorphic cerebral asymmetries in 2-deoxy-D-glucose uptake during postnatal development of the rat: correlations with age and relative brain activity. *Develop. Brain Res.* 3:341–347.
Ross, S. B., and Renyi, A. L. 1967a. Accumulation of tritiated 5-hydroxytryptamine in brain slice. *Life Sci.* 6:1407–15.
Ross, S. B., and Renyi, A. L. 1967b. Inhibition of the uptake of tritiated catecholamines by antidepressant and related agents. *Europ. J. Pharmacol.* 2:181–186.
Rossor, M., Garrett, N., and Iversen, L. 1980. No evidence for lateral asymmetry of neurotransmitters in post-mortem human brain. *J. Neurochem.* 35:743–745.
Rothman, A. H., and Glick, S. D. 1976. Differential effects of unilateral and bilateral caudate lesions on side preference and passive avoidance behavior in rats. *Brain Res.* 118:361–369.
Sackeim, H. A., Greenberg, M. S., Weiman, A. L., Gur, R. C., Hungerbuhler, J. P., and Geschwind, N. 1982. Hemispheric asymmetry in the expression of positive and negative emotions. *Arch. Neurol.* 39:210–218.
Satterfield, J. H., Lesser, L. L., Saul, R. E., and Cantwell, D. P. 1973. EEG aspects in the diagnosis and treatment of minimal brain dysfunction. *Ann. N.Y. Acad. Sci.* 205:274–282.
Schonfeld, A. R., and Glick, S. D. 1981. Handling-induced seizures and rotational behavior in the mongolian gerbil. *Pharmacol. Biochem. Behav.* 14:507–516.
Seeman, P. 1980. Brain dopamine receptors. *Pharmacol. Rev.* 32:229–313.
Shekim, W. O., Dekirmenjian, H., and Chapel, J. L. 1979. Urinary MHPG excretion in minimal brain dysfunction and its modification by d-amphetamine. *Am. J. Psychiatry* 136:667–671.

Sherman, G. F., Garbanati, J. A., Rosen, G. D., Yutzey, D. A., and Denenberg, V. H. 1980. Brain and behavioral asymmetries for spatial preference in rats. *Brain Res.* 192:61–67.
Snyder, S. H. 1976. The dopamine hypothesis of schizophrenia: focus on the dopamine receptor. *Am. J. Psychiatry* 133:197–202.
Snyder, S. H., and Coyle, J. T. 1969. Regional differences in ^{3}H-norepinephrine and ^{3}H-dopamine uptake into rat brain homogenates. *J. Pharmacol. Exp. Ther.* 165:78–86.
Sokoloff, L., Reivich, M., Kennedy, C., Des Rosiers, M. H., Patlak, C. S., Pettigrew, K. D., Sakurada, O., and Shinohara, M. 1977. The [^{14}C]deoxyglucose method for the measurement of local cerebral glucose utilization: theory, procedure and normal values in the conscious and anesthetized albino rat. *J. Neurochem.* 28:897–916.
Sokoloff, P., Martres, M. P., and Schwartz, J. C. 1980. ^{3}H-Apomorphine labels both dopamine postsynaptic receptors and autoreceptors. *Nature* 288:283–288.
Stein, L. and Wise, C. D. 1971. Possible etiology of schizophrenia: progressive damage to the noradrenergic reward system by 6-hydroxydopamine. *Science* 171:1032–36.
Stollack, J. S., and Glick, S. D. 1982. [^{3}H]Apomorphine binding to rat striatum reveals sexual and behavioral dimorphism. *Fed. Proc.* 41:1326.
Ungerstedt, U. 1971. Striatal dopamine release after amphetamine or nerve degeneration revealed by rotational behavior. *Acta Physiol. Scand. Suppl.* 367:49–68.
Valdes, J. J., Mactutus, C. F., Cory, R. N., and Cameron, W. R. 1981. Lateralization of norepinephrine, serotonin and choline uptake into hippocampal synaptosomes of sinistral rats. *Physiol. Behav.* 27:381–383.
vom Saal, F. S., and Bronson, F. H. 1978. In utero proximity of female mouse fetuses to males: effect on reproductive performance during later life. *Biol. Reprod.* 19:842–853.
vom Saal, F. S., and Bronson, F. H. 1980. Sexual characteristics of adult female mice are correlated with their blood testosterone levels during prenatal development. *Science* 208:597–599.
Wise, R. A., Spindler, J., deWit, H., and Gerber, G. J. 1978. Neuroleptic-induced "anhedonia" in rats: pimozide blocks reward quality of food. *Science* 210:262–264.
Witelson, S. F. 1976. Sex and the single hemisphere: specialization of the right hemisphere for spatial processing. *Science* 193:425–427.
Zimmerberg, B., and Glick, S. D. 1975. Changes in side preference during unilateral electrical stimulation of the caudate nucleus in rats. *Brain Res.* 86:335–338.
Zimmerberg, B., Glick, S. D., and Jerussi, T. P. 1974. Neurochemical correlate of a spatial preference in rats. *Science* 185:623–625.
Zimmerberg, B., Strumpf, A. J., and Glick, S. D. 1978. Cerebral asymmetry and left-right discrimination. *Brain Res.* 140:194–196.

Chapter 11

Lateralization of Neuroendocrine Control

Ida Gerendai

The possibility of cerebral lateralization for endocrine functions has only recently been raised. Experiments aimed at localizing extrahypothalamic structures critical for the control of hypothalamic hormone synthesis and release have not led to observations of cerebral lateralization because bilateral brain interventions were carried out at the same time. It was, furthermore, firmly believed that the endocrine target glands were controlled exclusively by hormonal mechanisms and that feedback control of the hypothalamus and pituitary was also purely hormonal.

For about a decade our group has been studying the existence and functional significance of direct neural connections between the hypothalamus and the peripheral endocrine glands — in particular, the ovary. Since the direct morphological demonstration of a long, probably multisynaptic pathway is virtually impossible with currently available tools, the problem had to be approached indirectly. With the compensatory growth of the intact organ following removal of the contralateral gland serving as the experimental model, we have been able to study the effect of removal of one of the paired endocrine glands on the cellular activity or specific hormone content of the two halves of the hypothalamus.

In another type of experiment we studied the effect of either a right- or left-sided lesion in the central nervous system on the compensatory hypertrophy of the remaining intact organ, which usually occurs after unilateral extirpation of a peripheral endocrine organ. We have also collected data on the lateralization of cerebral control of certain behavioral processes.

Gonadal Regulatory Mechanisms

Our early studies indicated that the protein-synthesizing activity of the arcuate nucleus of the hypothalamus differed significantly on the two sides after unilateral ovariectomy; contralateral to the hemigonadectomy, the arcuate neurons exhibited a more intense protein synthesis than those ipsilateral to the operation (Gerendai and Halasz, 1976). This asymmetry suggested the existence of direct neural connections between gonads and hypothalamus.

In order to test the possibility that asymmetrical changes following hemiovariectomy are related to lateralized alterations in levels of neurohormones, the luteinizing hormone-releasing hormone (LHRH) content of the two sides of the mediobasal hypothalamus (MBH) was measured two weeks after hemigonadectomy. We were surprised to find that in intact control females the right MBH contained significantly more LHRH than the left; this was the first demonstration of asymmetry in the endocrine hypothalamus (Gerendai et al., 1978).

Bilateral ovariectomy results in a drop in *total* LHRH content of the MBH (Wheaton and McCann, 1976), but we found that this is the result of a decrease in the LHRH level on the right side only— that is, on the side in which neurohormone content is higher in intact control females. Furthermore, unilateral ovariectomy resulted in a significant rise in LHRH in the half of the MBH ipsilateral to the operation. Thus, right-sided ovariectomy enhanced the LHRH asymmetry of the MBH present in normal controls, while left-sided ovariectomy diminished it (Fig. 11.1).

It is worth noting that in the adult male rat the two sides show no difference in neurohormone content of the MBH (Mizunuma et al., 1983). In spite of this symmetrical distribution, Nance et al. (1983) showed that the hemiorchidectomy-induced rise in follicle-stimulating hormone (FSH) could be blocked by unilateral hypothalamic deafferentation only when isolation of the right side of the hypothalamus was combined with right hemicastration. All of these data strongly suggest that the hypothalamic structures involved in neural control of gonadal function are lateralized. Biochemical asymmetry in the MBH is evident in females, but the lack of a similar bias in males does not exclude functional asymmetry. This view is supported by the findings of Glick et al. (1979), who demonstrated by use of the 2-deoxy-D-glucose technique metabolic asymmetry in favor of the right side in the rat hypothalamus. This is consistent with the higher LHRH content of the right MBH in female rats (Gerendai, 1978) and with the report of Nance and

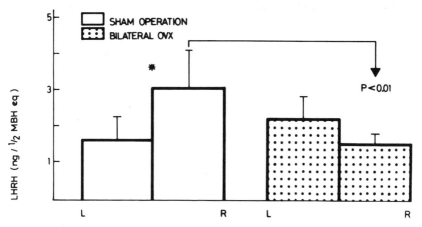

Fig. 11.1 Mean luteinizing hormone-releasing hormone (LHRH) content in the left (L) and right (R) side of the mediobasal hypothalamus in sham-operated animals and 2 weeks after bilateral ovariectomy (OVX). The LHRH content is expressed in nanograms per one-half of the MBH. The levels in the sham-operated animals did not differ significantly from those of controls.

Moger (1982) that structures critical for the hemicastration-induced serum FSH rise are located in the right hypothalamus.

The lateralization of extrahypothalamic structures participating in gonadal control has been demonstrated by studies in hemiovariectomized rats after a unilateral lesion of the locus coeruleus or unilateral vagotomy. Right-sided lesions of the locus coeruleus produced by kainic acid prevented the development of compensatory ovarian hypertrophy, regardless of the side of ovariectomy. In contrast, neurochemical lesions of the left locus coeruleus did not interfere with the usual compensatory ovarian growth. These findings indicate the involvement of the right, but not the left, locus coeruleus in the development of compensatory ovarian hypertrophy. It is not known whether the norepinephrine content of the locus coeruleus is symmetrical on the two sides. On the other hand, the distribution of norepinephrine in the human and the rat (Oke et al., 1978, 1980) thalamus is asymmetrical.

In adult female rats transection of the left or of the right vagus significantly reduced the rate of compensatory ovarian hypertrophy following hemispaying (Gerendai and Nemeskeri, 1983). In immature females operated on day 15, left-sided vagotomy and left- or right-sided ovariectomy did not lead to a compensatory ovarian hypertrophy as late as day 53. Furthermore, left-sided vagotomy, with or without unilateral ovariectomy, significantly delayed the onset of puberty, but right-sided vagotomy was without effect (Table 11.1). When the same operation was peformed on day 19, the effects were different. Several combinations of unilateral vagotomy and hemiovariectomy had no effect on either the rate of compensatory ovarian hypertrophy or the onset of puberty (that is, the findings in these animals were comparable to those in the controls). The one exception was right vagotomy plus right ovariectomy, which resulted in precocious puberty and enhanced compensatory ovarian hypertrophy.

The findings of our study indicate on the one hand the functional significance of the vagus in female gonadal control; on the other hand, they suggest that the asymmetrical control by the vagus changes during the prepubertal period from one side to the other, whereas there is no vagal asymmetry in control of compensatory ovarian hypertrophy in adulthood.

Age-related changes in hypothalamic laterality of control of compensatory ovarian hypertrophy have been demonstrated by Nance et al., 1983). In prepubertal female rats left-sided deafferentation of

Table 11.1 *Effect of unilateral vagotomy on compensatory ovarian hypertrophy in laboratory rats.*

Group vagotomized	Date of sacrifice after surgery	Right vagotomy		Left vagotomy	
		Right ovariectomy	Left ovariectomy	Right ovariectomy	Left ovariectomy
Adults	7 days	a	a	a	a
	48 days	b	a	b	b
Immatures	Operation day 15 after birth, sacrificed day 53	b	b	a	a
	Operation day 19 after birth, sacrificed day 53	c	b	b	d

a. Reduced compensatory ovarian hypertrophy.
b. No effect.
c. Enhanced compensatory ovarian hypertrophy.
d. Because of high mortality and retarded body growth, results were not comparable.

the hypothalamus blocked compensatory ovarian hypertrophy, while in adult females the same operation on the right side interfered with compensatory ovarian growth. These findings suggest that the dominant half of the hypothalamus and/or connections to and from each half of the hypothalamus alter between the pubertal and the adult period in females. The ontogenesis of hypothalamic laterality and of extrahypothalamic structures critical for the control of gonadotropin secretion needs further detailed study. The presence or absence of sex differences in the developmental pattern of hypothalamic and/or extrahypothalamic laterality must be specified. Finally, the relationship between asymmetry and the female gonadotropic cycle and the male tonic gonadotropic activity needs to be analyzed.

Sexually dimorphic asymmetries in the neonatal rat have been described by Ross and associates (1981). The incorporation of 2-deoxy-D-glucose into certain cerebral structures was asymmetric in females, but not in males. Furthermore, using the 2-deoxy-D-glucose method, Ross and colleagues (1982) demonstrated that both left-to-right and right-to-left maturational gradients exist in different brain areas during the postnatal developmental period in the rat. Such gradients are structure and sex dependent.

It is tempting to speculate whether this cerebral lateralization of gonadotropic control in rats could be responsible for the asymmetric ovarian activity seen in certain species. In the majority of mammals, including primates and laboratory rodents, the occurrence of ovulation is symmetrical or random, or alternates between the ovaries. In some mammals (a number of strains of bats, for instance) ovulation occurs predominantly from one ovary. Moreover, in birds, only the left ovary and oviduct become functional, whereas the right gonad is rudimentary.

Thyroid Control Mechanisms

In order to study the possible asymmetrical distribution of hypothalamic hormones involved in the control of peripheral endocrine glands, the thyrotropin-releasing hormone (TRH) content in the two halves of the MBH was measured in both intact and unilaterally or bilaterally thyroidectomized male rats (Gerendai et al., in preparation). In contrast to our previous observations of asymmetrical distribution of LHRH, we did not find lateral differences in the TRH content: the unilateral thyroidectomy-induced rise in TRH content was similar on the two sides. The neuroendocrine organization of the hypothalamo-hypophyseal-thyroid axis is thus not asymmetrical at the level of the MBH.

Nonetheless, an unequivocal asymmetry was found in extrahypothalamic pathways involved in thyroid function (Lewinski et al., 1982) when left- or right-sided separation of the hypothalamus from its caudal inputs was performed with and without hemithyroidectomy. Left-sided caudal deafferentation of the hypothalamus in animals with intact thyroids produced a decrease in basal mitotic activity of the follicular cells on both sides of the thyroid, whereas right-sided surgery was ineffective. In both left- and right-sided thyroidectomized animals, left-sided hypothalamic deafferentation prevented the hemithyroidectomy-induced rise in the mitotic index of the follicular cells of the remaining thyroid lobe. Right-sided hypothalamic deafferentation, on the other hand, did not interfere with the proliferation effect. It appears that the pathways responsible for the control of mitotic activity in the follicular cells of the thyroid gland lie predominantly on the left at the level of the diencephalic-mesencephalic junction.

The apparent discordance between the symmetrical distribution of hypothalamic TRH and the finding of an asymmetrical pathway for hypothalamic control of TRH secretion may be explained by the existence of bilateral projections from the dominant side. A pathway that is lateralized to one side but innervates both sides of the hypothalamus would explain symmetrical TRH content in the MBH, and it would also explain symmetrical changes of neurohormone content following unilateral thyroidectomy.

Thus far no data are available on TRH asymmetry in extrahypothalamic structures. TRH is known to act as a neurotransmitter in several cerebral and peripheral areas. Moreover, its administration has a beneficial effect in certain affective disorders. Although the hazards of extrapolating from animal experiments to human pathobiology are obvious, it is difficult to avoid speculating that a possible asymmetry in extrahypothalamic TRH distribution and/or endocrine activity of TRH-secreting cell groups may be relevant to the etiology and pathobiology of affective disorders.

Control of Prolactin Secretion

It is known that women undergoing unilateral mastectomy for breast carcinoma develop hyperprolactinemia, and in several cases galactorrhea as well (Herman et al., 1981). We have developed an animal model for mastectomy-induced hyperprolactinemia. Unilateral or bilateral mastectomies were performed in male rats. To our surprise, the ablation of all nipple areas and mammary glands on the right side induced hyperprolactinemia, whereas left-sided mastectomy was followed by hypoprolactinemia. In bilaterally mastectomized rats prolactin levels varied widely, with high levels in some animals and low levels in others. The average prolactin level of bilaterally mastectomized animals was, however, higher than that of the controls, but the rise was not significant. It is also noteworthy that the right mastectomy-induced hyperprolactinemia lasted for 2 weeks, while the decrease from left-sided mastectomy persisted for only 1 week (Gerendai, Prato, et al., submitted for publication).

These data indicate that pituitary prolactin secretion can be modified in opposite directions by left- or right-sided mastectomy. Since bilateral mastectomy results in the elevated plasma prolactin level characteristic of right-sided mastectomy, and the prolactin changes lasts longer when the operation is performed on the right side, it is possible that the right mammary gland and/or the neural pathways arising from it are dominant. Among adenohypophyseal hormones prolactin is unique in that its secretion is tonically inhibited by hypothalamic dopamine. Prolactin and growth hormone are distinguished by their lack of hormone-secreting peripheral target organs.

In order to explain the right and left mastectomy-induced differential prolactin responses, we proposed the existence of an asymmetrical pathway to the central nervous system arising in the mammary gland of male rats. This pathway would be analogous to the one in females, which is responsible for suckling-induced prolactin discharge. Whether or not asymmetry of function and underlying anatomical substrates are present in the female rat remains to be shown. The dopamine content in the tuberoinfundibular system of the two sides should also be studied. Finally, it would be useful to know whether, in the human, right or left mastectomy preferentially produces postoperative hyperprolactinemia and galactorrhea.

Cerebral Control of Grooming Behavior

Grooming behavior has been studied in unilaterally and bilaterally mastectomized rats. The presence of grooming behavior in a novel environment was recorded every 15 seconds over the course of 30 minutes (Glick et al., 1979). In animals mastectomized bilaterally and on the right side, grooming activity increased, whereas left mastectomy induced an insignificant decrease (Gerendai et al., 1984). Since published evidence suggests a positive correlation between prolactin levels and grooming activity, we conclude that the left and right mastectomy-induced alterations in grooming activity are not primarily the result of the existence of a lateralized control of grooming behavior, but rather the consequences of asymmetrical prolactin changes. We did not find a consistent relationship between prolactin level and grooming behavior in unilaterally vagotomized male rats. Left or right vagotomy produces a significant rise in plasma prolactin level (Gerendai et al., 1983). Right vagotomized animals, in spite of hyperprolactinemia, exhibited grooming activity similar to that of controls, whereas left-sided vagotomy produced hyperprolactinemia and excessive grooming (Gerendai et al., 1984). We assume, therefore, that one of the mechanisms of grooming behavior is prolactin sensitive, while another is independent of prolactin and is lateralized.

Prolactin is known to interfere with dopamine transmission in the hypothalamus (Neill, 1980) and in the substantia nigra (Lichtensteiger and Lienhart, 1975). Changes in dopamine turnover induced by prolactin have been reported to occur in the nucleus accumbens and the medial caudate (Fuxe et al., 1977, 1978), and in the striatum (Perkins and Westfall, 1978). Furthermore, the substantia nigra has been implicated as the site of prolactin action on grooming behavior (Drago et al., 1981).

The lateral asymmetry of the nigrostriatal pathway of the rat, and its functional significance in spatial preferences, have been reviewed by Glick and Ross (1981). The vagotomy-induced changes in grooming behavior may reflect the existence of a lateralized mechanism. This view is supported by the finding of asymmetry in paw-licking response in unilaterally vagotomized animals. The latency of paw-licking response on a hot-plate test did not differ in right-sided vagotomized rats from that of controls. Whereas left vagotomy showed a higher level of paw-licking responses than sham-operated or intact control rats (Gerendai et al., 1984). Vagal fibers, or structures receiving vagal inputs, which are involved in the perception of pain, apparently are left-biased. Taking into account all the data obtained in unilaterally vagotomized rats, we conclude that the lateralized function of the vagus nerve depends on sex and age as well as on the particular function studied.

Mortality Rates in Brain Interventions

In order to study neuronal mechanisms underlying compensatory ovarian hypertrophy after hemicastration, lesions were placed in different regions of the central nervous system. By chance we noticed that the majority of surviving animals undergoing brain surgery came from the group in which the lesion had been placed on the left side. This empirical observation led us to carry out a retrospective analysis of the postoperative mortality of animals having left-sided or right-sided brain intervention (Table 11.2). The following operations were included: unilateral lesions of the medial preoptic area with kainic acid; unilateral frontal deafferentation of the MBH; unilateral complete deafferentation of the MBH; unilateral lesions of the hypothalamic arcuate region with kainic acid; unilateral transection of the diencephalic-mesencephalic border; unilateral lesions of the locus coeruleus with kainic acid; and unilateral lesions of the medial raphe nucleus with kainic acid. The postoperative period lasted 1 to 2 weeks. Surprisingly, in each brain region right-sided surgery caused a much higher mortality rate than identical lesions on the left side. The percentage difference between right-sided and left-sided brain interventions was highest in animals undergoing MBH surgery, and lowest in animals lesioned in the medial preoptic region.

In spite of the unequivocal results, it is difficult to explain why right-sided surgery induces a mortality rate more than twice that of left-sided surgery. It is tempting to hypothesize that the right side of the brain has greater relevance to vital functions. Robinson and Coyle (1980) reported that ligation of the right middle cerebral artery resulted in spontaneous hyperactivity and bilateral decrease of catecholamine content in the brain, whereas ligation of the left middle cerebral artery did not produce similar alterations.

Table 11.2 Mortality rate in laboratory rats following right-sided or left-sided brain surgery.

	Mortality in right-sided surgery		Mortality in left-sided surgery		
Site of surgery	No.[a]	%	No.	%	p
Medial preoptic area lesion	25/57	45.6	14/53	26.4	b
Medial basal hypothalamus deafferentation or lesion	27/50	54.0	8/51	15.7	c
Lower brain-stem intervention	23/55	41.8	5/39	12.8	d
Total	76/162	46.9	27/143	18.9	c

[a] Number of animals dying after surgery/number of animals operated on.
b < .05.
c < .0005.
d < .005.

Analysis of the data presented above suggests that cerebral lateralization may play a role in many systems, including endocrine regulatory mechanisms. Our sporadic observations indicate the need for additional systematic investigations. Studies should be extended to include other parts of the endocrine system (for instance, there are no data on lateralization of control either of adrenal secretions or of growth hormone). Furthermore, it is imperative that careful studies be done on the ontogeny and probable sexual dimorphism of the functionally and anatomically asymmetrical endocrine control systems. Such investigations might clarify whether functional endocrine asymmetries are the consequences of morphological or biochemical asymmetry, or of both.

Most of the data presented above are from lesion experiments. It is important that experiments be performed using stimulation of intact neuroendocrine structures. We might thereby fill the gap that exists in our understanding of lateralization of endocrine control as well as obtain further data on the fundamental biological significance of brain lateralization. It must be emphasized that the source of our information on endocrine lateralization has been exclusively the laboratory rat, an animal model that might serve as a starting point for the study of endocrine lateralization in other species. A critical question, to which we have no answer at present, is whether similar lateralization of endocrine control exists in man. Direct studies in the human are difficult; however, clinical observations stimulated by the data in this chapter may produce evidence of lateralization of endocrine symptoms in human disease, and could in turn lead to new findings that will stimulate research in the animal laboratory. The finding of lateralization in endocrine systems, finally, prompts a search for asymmetry in other bodily systems.

References

Drago, F., Bohus, B., Canonico, P. L., and Scapagnini, U. 1981. Prolactin induces grooming in the rat: possible involvement of nigrostriatal dopaminergic system. *Pharmcol. Biochem. Behav.* 15:61–63.

Fuxe, K., Eneroth, P., Gustaffson, J. A., Lofstrom, A. and Skett, P. 1977. Dopamine in the nucleus accumbens: preferential increase of DA turnover by rat prolactin. *Brain Res.* 122:177–182.

Fuxe, K., Andersson, K., Hokfelt, T., and Agnati, L. F. 1978. Prolactin-monoamine interactions in the rat brain and their importance in regulation of LH and prolactin secretion. In C. Robin and M. Harter, eds., *Progress in Prolactin Physiology and Pathology*. Amsterdam: Elsevier, pp. 95–109.

Gerendai, I., and Halasz, B. 1976. Hemigonadectomy-induced changes in the protein-synthesizing activity of the rat hypothalamic arcuate nucleus. *Neurendocrinology* 21:331-337.

Gerendai, I., and Nemeskeri, A. 1983. The effect of unilateral vagotomy on compensatory ovarian hypertrophy and on the onset of puberty. In E. Endroczi, L. Angelluci, U. Scapagnini, and D. deWied, eds., *Neuropeptides, Neurotransmitters, and Regulation of Endocrine Processes*. Budapest: Akademiai Kiado, pp. 191-198.

Gerendai, I., Rotsztejn, W., Marchetti, B., Kordon, C., and Scapagnini, U. 1978. Unilateral ovariectomy-induced luteinizing hormone-releasing hormone content changes in the two halves of the mediobasal hypothalamus. *Neurosci. Lett.* 9:333-336.

Gerendai, I., Clementi, G., Prato, A., and Scapagnini, U. 1983. Unilateral vagotomy induces hyperprolactinemia in male rats. *Neuroendocrine Lett.* 5:41-45.

Gerendai, I., Drago, F., Continella, G., and Scapagnini, U. 1984. Effects of mastectomy and vagotomy on grooming behavior of the rat: possible involvement of prolactin. *Physiol. Behav.*: in press.

Gerendai, I., Prato, A., Clementi, G., and Scapagnini, U. Effect of unilateral or bilateral mastectomy on prolactin secretion in male rats. Submitted for publication.

Gerendai, I., Nemeskeri, A., Faivre-Bauman, A., Grouselle, D., and Tixier-Vidal, A. Effect of unilateral bilateral thyroidectomy on the TRH content of the hypothalamus of the two sides. In preparation.

Gispen, W. H., Wiegant, V. M., Greven, H. M., and deWied, D. 1975. The induction of excessive grooming in the rat by intraventricular application of peptides derived from ACTH: structure-activity studies. *Life Sci.* 17:645-652.

Glick, S. D., and Ross, D. A. 1981. Lateralization of function in the rat brain. Basic mechanisms may be operative in humans. *TINS* 4:196-199.

Glick, S. D., Meibach,, R. C., Cox, R. D., and Maayani, S. 1979. Multiple and interrelated functional asymmetries in rat brain. *Life Sci.* 25:395-400.

Herman, V., Kalk, W. J., de Moor, N. G., and Levin, J. 1981. Serum prolactin after chest wall surgery: elevated levels after mastectomy. *J. Clin. Endocrinol. Metab.* 52:148-153.

Lewinski, A., Gerendai, I., Pawlikowski, M., and Halasz, B. 1982. Unilateral posterior deafferentation of the hypothalamus and mitotic activity of thyroid follicular cells under normal conditions and after hemithyroidectomy. *Endocrinol. Exp.* 16:75-80.

Lichtensteiger, W., and Lienhart, R. 1975. Central action of α-MSH and prolactin: simultaneous responses of hypothalamic and mesencephalic dopamine systems. In E. Endroczi, ed., *Cellular and Molecular Bases of Neuroendocrine Processes*. Budapest: Akademiai Kiado, pp. 211-221.

Mizunuma, H., DePalatis, L. R., and McCann, S. M. 1983. Effect of unilateral orchidectomy on plasma FSH concentration: evidence for a direct neural connection between testes and CNS. *Neuroendocrinology* 37:291–296.

Nance, D. M., and Moger, W. H. 1982. Ipsilateral hypothalamic deafferentation blocks the increase in serum FSH following hemicastration. *Brain Res. Bull.* 8:299–302.

Nance, D. M., White, J. P., and Moger, W. H. 1983. Neural regulation of the ovary: evidence for hypothalamic asymmetry in endocrine control. *Brain Res. Bull.* 10:353–355.

Neill, J. D. 1980. Neuroendocrine regulation of prolactin secretion. In L. Martini and W. F. Ganong, eds., *Frontiers in Neuroendocrinology*. New York: Raven Press, vol. 6, pp. 129–155.

Oke, A., Keller, R., Mefford, I., and Adams, R. N. 1978. Lateralization of norepinephrine in the human thalamus. *Science* 200:1411–13.

Oke, A., Lewis, R., and Adams, R. N. 1980. Hemispheric asymmetry of norepinephrine distribution in rat thalamus. *Brain Res.* 188:269–272.

Perkins, N. A., and Westfall, T. C. 1978. The effect of prolactin on dopamine release from rat striatum and medial basal hypothalamus. *Neurosciences* 3:59–63.

Robinson, R. G., and Coyle, J. T. 1980. The differential effect of right versus left hemispheric cerebral infarction on catecholamines and behavior in the rat. *Brain Res.* 188:63–78.

Ross, D. A., Glick, S. D., and Meibach, R. C. 1981. Sexually dimorphic brain and behavioral asymmetries in the neonatal rat. *Proc. Natl. Acad. Sci. USA* 78:1958–61.

Ross, D. A., Glick, S. D., and Meibach, R. C. 1982. Sexually dimorphic cerebral asymmetries in 2-deoxy-D-glucose uptake during postnatal development of the rat: correlations with age and relative brain activity. *Develop. Brain Res.* 3:341–347.

Wheaton, J. E., and McCann, S. M. 1976. Luteinizing hormone-releasing hormone in peripheral plasma and hypothalamus of normal and ovariectomized rats. *Neuroendocrinology* 20:296–310.

Chapter 12
Experimental Modification of Gyral Patterns

Patricia S. Goldman-Rakic
Pasko Rakic

The formation of cerebral convolutions is better known than understood, more talked about than studied. Although the basic anatomical description of adult human cerebral sulcal and gyral configurations and the approximate timetables of their ontogenesis were worked out during the last century, the mechanisms by which a mass of cortical neurons shifts in position, folds, and undergoes differential growth to produce a cerebral surface with its characteristic convoluted form are still obscure. Most of our knowledge is based on observations of normal adult and developing human brains and on reports of cases with various pathological conditions that produce or are the result of dysgenesis of the cortical surface.

Experimental data on the genesis of convolutions and the emergence of cortical parcellation are lacking for several reasons. First, experimental neuroembryological research on the neocortex was traditionally carried out on lissencephalic (that is, not convoluted) rodent brains. Second, since convolutions are mostly in place by the time of birth, analysis of the mechanisms underlying development of cortical folding in gyrencephalic (convoluted) brains requires experimental manipulation of the cerebrum before birth, a period in which the fetus has been inaccessible to complex surgical procedures. However, recent advances in techniques of prenatal neurosurgery, which enable precise and extensive removal of selected cortical areas at critical fetal ages, have opened up new responsibilities for the study of this important developmental process (Rakic and Goldman-Rakic, 1983).

In this chapter we outline the normal timetable of corticogenesis, and the mode and sequence of development of sulci and gyri in rhesus monkey and human cerebri. We describe ongoing experiments in our laboratories that are aimed at determining the possible role of neuronal connections and their competition in establishing cytoarchitectonic parcellation of the neocortex and the pattern of fissuration in the hemispheric surface in primates. And, finally, we draw some parallels between the development of sulci and gyri in rhesus monkey and human, and discuss possible implications of our experiments for the pathogenesis of neonatal cerebral injury and the development of hemispheric asymmetry.

Normal Corticogenesis and Convolutional Development

The emergence of cytoarchitectonic and functionally distinct cortical subdivisions marks the crowning achievement of cerebral development. In the primate brain major cytoarchitectonic areas tend to be separated by fissures, whereas finer subdivisions are framed by primary and secondary sulci (Smith, 1906). This very basic fact indicates that the emergence of sulci and gyri may provide some general guidelines on the timing, sequence, and tempo of cytoarchitectonic differentiation of the cerebral surface, and on the factors involved in the process. It is remarkable how little is known about the influences that govern the parcellation of the human cerebral cortex. Although in the past the significance of neocortical cytoarchitectonic subdivisions was a subject of considerable controversy and heated arguments took place about their precise boundaries, it is now basically accepted that the 80,000 sq mm surface of the cortex is discontinuous in its topographic organization and is locally specialized in function (see Schmitt et al., 1981; Rakic and Goldman-Rakic, 1982).

Modern methods of anatomy and physiology have shown that each cytoarchitectonic area has specific intrinsic and extrinsic connections with other areas and with various subcortical structures (Pandya and Sanides, 1973). It is also well established that the gross size of cortical regions and the extent of cytoarchitectonic fields in the human brain show significant variability and that there are consistent left-right asymmetries (Geschwind and Levitsky, 1968; Galaburda et al., 1978). Yet we do not know when and how specific regional differences are generated. Are the functions of cortical neurons that subserve one or another modality determined before neurons migrate to their final destinations, or after afferent input invades the developing cortical plate? What influence do specific afferents have in the genesis of cytoarchitectonic diversity, and do they play a role in determining the sizes of individual fields? Is the pattern of connectivity related to the pattern of convolutions? We are convinced that a comparison of the voluminous data on the pathology of the human brain with new information obtained by modern anatomical and cytological methods from research on the monkey brain may help to resolve some of these important issues.

In the midgestational period the cerebral hemispheres of both human and monkey are basically smooth and resemble the cerebrum of adult lissencephalic mammals. As illustrated in Fig. 12.1, only a shallow sylvian fissure, a central sulcus, and the calcarine

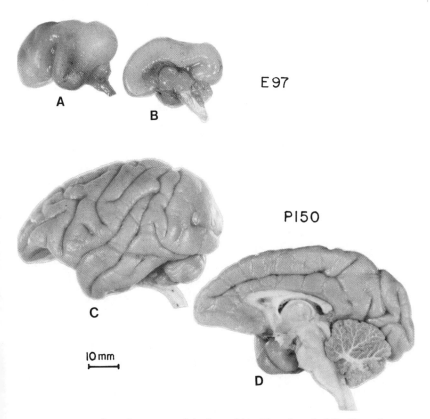

Fig. 12.1 *External configuration of the lateral (A, C) and medial (B, D) surfaces of the monkey cerebrum at the ninety-seventh embryonic day (E97) and hundred fiftieth postnatal day (P150). Both brains were fixed identically by vascular perfusion with a glutaraldehyde-paraformaldehyde mixture in phosphate buffer, taken from the skull one hour later and photographed at the same magnification. The increase in the neocortical surface and the development of convolutions occur after virtually all cortical neurons have been generated.*

fissure are evident in the rhesus monkey at the beginning of the second half of gestation. Secondary fissures develop later in gestation, and the tertiary fissures emerge only within the last fetal month in both man and monkey. Likewise, during the 40-week period of human gestation, gyri develop most rapidly between fetal weeks 26 and 35 (Connolly, 1940; Chi et al., 1977).

It is not our purpose to provide here a detailed anatomical account and precise timetable of convolutional development. We wish only to emphasize two points. First, the process of gyration involves a severalfold increase in surface area, since approximately two-thirds of the neocortex of the adult brain is buried within the depths of sulci and fissures. Second, although the times of emergence of fissures and sulci are readily determined, sulci or fissures form *only* after all cortical neurons have been generated. We have done detailed autoradiographic studies on monkeys in which [^{3}H]thymidine was used to label permanently the nuclear DNA of cortical neurons that undergo their last cell division at the time of injection of the isotope. These studies show that all neocortical neurons in the rhesus monkey are produced by midgestation (Rakic, 1974, 1975, 1976a, 1982), prior to the formation of convolutions.

For obvious reasons equivalent experimental thymidine labeling data are not available for human neurogenesis. However, extrapolations from information available on times of cell origins in monkey to the timing of various cellular events in humans (using as an aid several morphological criteria of maturation) are concordant with the view that production of neocortical cells in humans is also completed by or before fetal week 20 — in other words, in the first half of pregnancy (Sidman and Rakic, 1973, 1982; Rakic, 1978). Thus, formation of the convolutional pattern of the primate telencephalon is basically a postmigratory phenomenon.

This is not to say, however, that severe malformations of convolutional patterns cannot be produced by disturbances of migration. Rather, the fact that the process of migration is completed while the hemispheric surface is still basically smooth suggests that the normal development of fissures, sulci, and gyri may be difficult to explain primarily in simple mechanical terms — for example, as the outcome of the accumulation of late-generated cells within superficial cortical layers which receive the last complement of neurons. Data on the timing of neuronal origin and of formation of sulci in the monkey also indicate that the local variations in cell proliferation rates that have been suggested as a cause of fissuration of the cerebellum (Mares and Lodin, 1970) cannot play a significant role in the primate cerebrum. Nor can one explain buckling of the neocortical surface simply as an outcome of expansive growth within a limited cranial space (Papez, 1929; Clark, 1945), as was argued by Young two decades ago (1962). Experiments involving decompression of intracranial space by the partial removal of the skull in fetal sheep did not result in a grossly diminished or altered pattern of fissuration (Barron, 1950). More recent and sophisticated theories (Richman et al., 1975; Todd, 1982) deal with mechanisms of folding, but fail to explain the specificity in the shape and size of convolutional patterns. Likewise, the appearance of fissures, sulci, and gyri do not merely result from an increase in quantity of gray matter relative to white matter (Smith, 1931). Only a systematic and detailed analysis of all relevant hodological and developmental mechanisms associated with the formation of the cortex can identify the ontogenetic and phylogenetic forces that produce convolutions in higher mammalian species.

Experimentally Altered Convolutional Development

The possibility of using prenatal neurosurgery with a high fetal survival rate to remove well-delineated regions of the cerebral cortex enables a more systematic approach to this unsolved problem in developmental neurobiology. In the course of our studies on the mechanism of prenatal development of brain connections and the behavioral consequences of prenatal cerebral damage, we have had the opportunity to examine the brains of a number of monkeys who underwent either unilateral or bilateral resections of the prospective prefrontal, sensory, motor, or primary visual cortex at various gestational ages, and to compare these with brains of normal animals or those that underwent comparable neurosurgical procedures at selected postnatal ages and as adults. (Details of the technique and potential advantages of prenatal surgery for the study of mechanisms of brain development in large mammals are given in Rakic and Goldman-Rakic, 1983.)

We found that the structural consequences of direct or indirect injury to groups of immature nerve cells can take a wide variety of forms, ranging from gross morphological distortions of the convolutional pattern to changes in long-tract projections and cellular and subcellular modifications of synaptic contacts and cytoplasmic organelles. The simplest and most obvious change encountered was the dramatic alteration of the cerebral surface. Even cursory examination of the brains of prenatally operated monkeys revealed an abnormal configuration of sulci and gyri that was not a simple or direct consequence of surgery and/or removal of cortical tissue. Abnormal sulci and gyri had formed not only in areas bordering the lesion, but also in remote cortical regions. Examples of these anomalies are shown in Fig. 12.2, which compares lateral-view reconstructions of a normal brain with comparable reconstruction from the brain of a monkey that had both frontal lobes resected at E106 (106th embryonic day). Unusual sulci were found, for example, on the dorsolateral surface of the occipital lobe, a region that is normally smooth. Also, at the superior margin of the parietal lobe, the lunate sulcus bifurcated on the lateral surface in an unusual pattern in each hemisphere. Abnormal sulci appeared on both sides, and while not fully equivalent in size, seemed roughly homologous. Although in the two-dimensional drawing of Fig. 12.2 the abnormal sulci are indicated by short lines, examination of the transverse sections on which the reconstruction is based reveals that they are often deep invaginations of cortex buried beneath the surface of the cerebrum (Goldman and Galkin, 1978). These changes are not caused by disturbance in blood flow, since sham operations with coagulation of the surface arteries and veins did not produce a similar effect. Also the damage to fibroblasts and basal laminae that has been postulated to play a local role in cerebellar folding (Sievers et al., 1981) is unlikely to produce convolutional changes at sites remote from the lesion, as in our material.

Similarly dramatic, though geometrically different, changes can be obtained following partial removal of the primary visual cortex (area 17 of Brodmann) in fetal monkeys. This series of experiments was initiated to analyze the effect of visual cortical lesions on the lateral geniculate nucleus of the thalamus and the transneuronal effects on the ganglion cells in the retina. In a case in which the prospective primary visual cortex had been resected at E83, and the fetus allowed to survive to 1½ years of age, the cortical surface was most dramatically altered (Fig. 12.3). On the side of the lesion the occipital pole with its usually smooth lateral surface was missing, and only a relatively small, shrunken vestige of cortex was found at the site. This was expected, since the prenatal resection of occipital lobe was extensive. As a consequence of the large cortical lesion, most of the lateral geniculate nucleus showed retrograde degeneration (Ogren et al., 1983). However, in front of the lesion an area roughly equivalent to the so-called inferior parietal lobule (Brodmann's area 7) was altered and significantly larger than on the unoperated side. This cortical area is delineated anteriorly by the intraparietal sulcus (IP) and posteriorly by the lunate sulcus (L). Remarkably, the shape of the inferior parietal lobule seemed to be well

Experimental Modification of Gyral Patterns 185

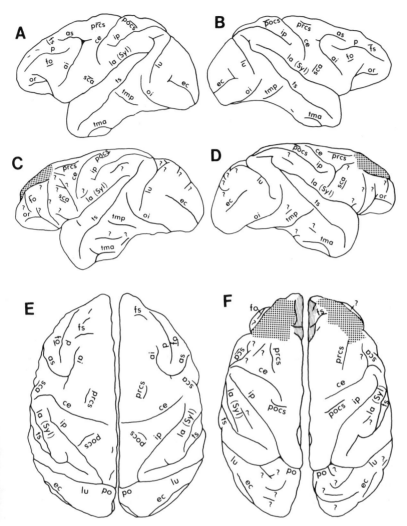

Fig. 12.2 Lateral and dorsal views of the cerebral hemispheres in a normal unoperated 2½-year-old monkey (A, B, E) and in a monkey subjected to frontal lobe resection at fetal day 106, returned to the uterus, and allowed to survive for 2½ years (C, D, F). The stippled surface indicates lesioned areas. Note the preponderance of ectopic sulci (indicated by question marks) in the prefrontal, temporal, and occipital lobes. Ai, inferior ramus of arcuate sulcus; as, superior ramus of arcuate sulcus; ce, central sulcus; ec, ectocalcarine sulcus; fo, orbital-frontal sulcus; fs, sulcus frontalis superior; ip, intraparietal sulcus; la (syl), lateral (sylvian) fissure; lu, lunate sulcus; oi, inferior occipital sulcus; po, parieto-occipital incisure; pocs, superior postcentral sulcus; pros, superior precentral dimple; sca, anterior subcentral dimple; tma, anterior middle temporal sulcus; tmp, posterior middle temporal sulcus; ts, superior temporal sulcus. (After Goldman and Galkin, 1978; by permission of Elsevier Biomedical Press.)

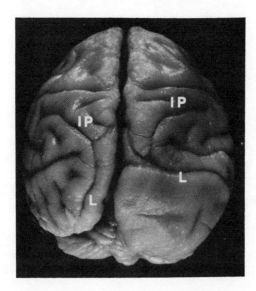

Fig. 12.3 Dorsal surface of the cerebrum of a rhesus monkey whose left occipital lobe was removed at day 83 of gestation (E83). This photograph illustrates the dramatic reorganization of fissures that can occur following experimental cortical resection. The inferior parietal lobule, bounded by the intraparietal (IP) sulcus anteriorly and the lunate (L) sulcus posteriorly, occupies nearly twice the surface area in the resected hemisphere that it does in the intact hemisphere.

preserved and easily recognizable in spite of changes in the angle of the IP and L sulci and its substantially larger size. Although there was an enlargement of the lateral ventricle in the temporal lobe on the lesioned side, the ventricular size within the parietal lobe was normal in volume and about equal on both sides. Furthermore, the sulci of the parietal lobe were as deep on the lesioned side as on the unlesioned side. Thus, the increase in the surface area of the inferior parietal lobule (as defined by the intraparietal and lunate sulci) on the lesioned side was a real phenomenon. Detailed cytoarchitectonic morphometric analysis of this alteration has not yet been completed.

One important difference in the effects of lesions of the frontal and occipital poles is of considerable significance for a discussion of possible mechanisms of fissuration. Lesions of the frontal lobe produce mainly bilateral changes in sulcal pattern (Fig. 12.2), while lesions of the occipital lobe produce grossly asymmetrical changes (Fig. 12.3). This difference in lesion effect may be due to the fact that callosal connections are prominent in the frontal and absent in the primary visual areas.

Possible Mechanisms of Convolutional Development

It was in the course of these experiments using prenatal neurosurgery that we postulated that long-tract connectivity and axonal competition for synaptic space in the developing cortical plate played an important role. It became apparent that the timetable of development of convolutions in the rhesus monkey brain coincides roughly with the time interval over which there is an influx of thalamic and corticocortical afferents into the cortex (Goldman and Galkin, 1978; Goldman-Rakic, 1981). Furthermore, studies on development of visual connections in monkeys show that the thalamocortical innervation of the occipital lobe occurs mostly during the third quarter of gestation (Rakic, 1976b, 1977, 1979). Other thalamic connections in the primate brain, for example those to prefrontal areas, appear to invade their cortical targets over the same period of time (Goldman and Galkin, 1978). Intrahemispheric and callosal fibers innervate the cortex somewhat later — roughly between E124 and E150 (Goldman-Rakic, 1981; Schwartz and Goldman-Rakic, 1982). This cerebral fissuration begins at about the same time that thalamocortical afferents invade the cortex, and it assumes its mature pattern during the major ingrowth of corticocortical connections. It may not be coincidental that anomalous features in the sulcal pattern can be experimentally induced only in the brains of monkeys that have been operated on before the end of this period.

The correspondence between the timetables of fissure development and of innervation of the cortical plate suggests that these two major developmental events are causally related. A temporal correlation between these events is not in itself proof of a causal relation; nonetheless, this hypothesis may explain how the disruption of a rather small part of the cortex can produce widespread changes that encompass the entire cerebral surface of both hemispheres. If neurons are removed before their axons have reached their ipsilateral cortical targets in adjacent and distant destinations, these target structures will be deprived of a considerable portion of their normal input. They would be subjected to unusual structural forces resulting from abnormal numbers and arrangements of ingrowing fibers. Furthermore, since it is known that transneuronal degeneration occurs more extensively in immature than in mature brains (see Cowan, 1970), it may be expected that neurons deprived of their normal inputs in the cortical target regions would degenerate in greater numbers and proportions in fetal than in more mature animals. Finally, this effect could be extended to callosal neurons to corresponding loci in the opposite hemisphere, which would also degenerate or become rearranged, so that bilateral changes in homotopic cortical zones would be seen.

According to this model, protection from transneuronal degenerative processes occurs with the successful collateralization of afferent and efferent systems, and thus would not occur to any noticeable extent after connectivity in the brain is well advanced (presumably several weeks before birth). The model explains why subsequent to prefrontal or visual cortical lesions, anomalous sulci and gyri were not observed in the ipsilateral somatosensory or primary motor cortexes, regions that do not have direct connections with prefrontal and visual cortices. Finally, the model predicts that lesions in areas that normally do not have callosal connections would not produce bilateral changes in surface configuration. Indeed, only unilateral changes were observed after a lesion of the primary visual area, which lacks callosal connections.

The proposed hypothesis assumes that various classes of afferents compete for appropriate synaptic territories in the developing cortical plate. This competition establishes the final pattern of distribution and density of connections, specific for each cortical field. Their relative volume and density may cause buckling of the surface; indeed, gyri appear to receive strong projections from the thalamus, while sulci receive weak thalamic projections. Furthermore, major gyri tend to receive projections from distant body regions (Welker and Campos, 1963); conversely, the depths of sulci seem to have the highest density and overlap of corticocortical connections (Goldman-Rakic and Schwartz, 1982). The innervation of the brain stem monoamine input to the cortex is more dense in the sulcal invaginations than at the summit of gyri (Levitt et al., 1982). We postulate that genetically or environmentally induced changes in this regional assortment of inputs can cause changes in convolutional pattern.

The present results do not refute the hypothesis that earlier-occurring migration defects, which prevent neurons from arriving at appropriate positions within the laminae of the developing cortical plate, may play an important role in the pathogenesis of some convolutional malformations. Among these migrational alternations are lissencephaly, pachygyria, and certain types of polymicrogyria (see, for instance, Stewart et al., 1975; Ervard et al., 1978; Volpe, 1981). However, our results indicate that there may be another class of cortical malformation that is induced at later developmental stages. Some cases of polymicrogyria and distortions of the cerebral surface occur after migration has been completed (Volpe, 1981). Such examples include cases of laminar necrosis due to external agents (such as carbon monoxide poisoning) that affect some deep layer cells but predominantly destroy projection neurons of layer III. Significantly, layer III is the major source of ipsilateral and contralateral corticocortical connections in rhesus monkeys (Schwartz and Goldman-Rakic, 1983) and presumably in humans. Thus, at least some classes of convolutional abnormality in man can be caused by selective elimination of corticocortical connections.

Our data on the development of convolutions are by-products of ongoing investigations in our laboratories that are directed toward different goals. We have not yet carried out all the controls and additional experiments necessary to test the proposed hypothesis rigorously. For example, enlargement of the parietal area delineated by the lunate and intraparietal sulci may conceivably be caused by one or more developmental mechanisms about which we can only speculate. It is possible that be removing area 17 we have deafferented the prestriate visual areas (Brodmann's 18, 19, and 21), which are then invaded by afferents normally restricted to area 17. If this is so, then both the shape and the size of area 17 may be determined by its input. The pattern of this input, in turn, may be influenced by competition with afferents from adjacent cortical areas. Such a possibility has important conceptual and practical importance and can be tested by presently available experimental anatomical methods.

Another possible mechanism that may be partially or wholly responsible for the unilateral or bilateral enlargement of a given cytoarchitectonic field is a reduced rate of cell death within a circumscribed region of the developing cortical plate. The opposite possibility — namely, that there is an accretion of new neurons — has been eliminated by studies of injections of [^{3}H]thymidine following fetal cortical resection (Goldman-Rakic, 1980). A diminished rate of naturally occurring cell elimination has been observed in other systems of the primate brain as a result of lesions in distant but synaptically related structures. For example, enucleation of one eye at critical fetal stages in rhesus monkeys results in a significantly larger number of optic axons and presumably higher survival rate of retinal ganglion cells in the intact eye (Rakic and Riley, 1983). This finding illustrates how one population of neurons can control the number of neurons in another structure with which it shares a common synaptic territory. One is tempted to speculate that similar competitive mechanisms may underlie compensatory enlargement, or possibly even the enhancement of function, of one cortical region when another is damaged during critical stages of development. Obviously, the next step is to analyze neurophysiologically and behaviorally the function of abnormally enlarged cytoarchitectonic areas following selected cortical lesions at fetal ages.

The proposed hypothesis that the pattern of connectivity and competition between various inputs for synaptic space available in the developing cortical plate may play a role in shaping architectonic fields, and secondarily in sculpting the cerebral surface, has obvious implications for understanding asymmetry of the cerebral hemispheres in man. The size of a given cytoarchitectonic field may be determined by the volume of the input originating either from the thalamus or from other cortical areas on the same or the opposite side. There is evidence that asymmetry of the cortical fields in the human brain occurs in parallel with asymmetry in thalamic nuclei that project to these fields (Eidelberg and Galaburda, 1982). Likewise, evidence from experimental and genetically altered brain connectivity in rodents indicates that peripheral input to the cortex can influence the extent and pattern of topographical representation on the cortical surface (Van der Loos, 1979). The genetic factors which determine the number of neurons that send projections to the cortex or the pathological factors that act on the different axonal inputs could ultimately result in unequal surfaces on the two sides of the brain. It should be emphasized that these major and obvious asymmetries are not different in kind from less conspicuous variations produced by constant small fluctuations in the environment. These environmentally determined variations may play a major role in determining the unique features of each individual.

References

Barron, D. H. 1950. An experimental analysis of some factors involved in the development of fissure pattern in the cerebral cortex. *J. Exp. Zool.* 113:553–573.

Chi, J. G., Dooling, E., and Gilles, F. H. 1977. Gyral development of the human brain. *Ann. Neurol.* 1:86–93.

Clark, W. E. LeG. 1945. Deformation patterns in the cerebral cortex. In W. E. LeG. Clark and P. B. Medawar, eds., *Essays on Growth and Form Presented to D'Arcy Wentworth Thompson*. Oxford: Clarendon Press, pp. 1–22.

Connolly, C. J. 1940. Development of cerebral sulci. *Am. J. Phys. Anthropol.* 26:113–149.

Cowan, W. M. 1970. Anterograde and retrograde transneuronal degeneration in the central and peripheral nervous system. In W. J. H. Nauta and S. O. E. Ebbesson, eds., *Contemporary Research Methods in Neuroanatomy*. New York: Springer-Verlag, pp. 217–251.

Eidelberg, D., and Galaburda, A. M. 1982. Symmetry and asymmetry in the human posterior thalamus. *Arch. Neurol.* 39:325–332.

Ervard, P., Caviness, V. S., Prats-Vinas, J., and Lyon, G. 1978. The mechanism of arrest of neuronal migration in the Zellweger malformation: an hypothesis based upon cytoarchitectonic analysis. *Acta Neuropathol.* 41:109–117.

Galaburda, A. M., LeMay, M., Kemper, T. L., and Geschwind, N. 1978. Right-left asymmetries in the brain. *Science* 199:852–856.

Geschwind, N., and Levitsky, W. 1968. Human brain: left-right asymmetries in temporal speech region. *Science* 161:167–168.
Goldman, P. S., and Galkin, T. W. 1978. Prenatal removal of frontal association cortex in the rhesus monkey: anatomical and functional consequences in postnatal life. *Brain Res.* 52:451–485.
Goldman-Rakic, P. S. 1980. Morphological consequence of prenatal injury to the primate brain. *Prog. Brain Res.* 53:3–19.
Goldman-Rakic, P. S. 1981. Development and plasticity of primate frontal association cortex. In F. O. Schmitt, ed., *The Organization of the Cerebral Cortex*. Cambridge, Massachusetts: MIT Press, pp. 69–97.
Goldman-Rakic, P. S., and Schwartz, M. E. 1982. Interdigitation of contralateral and ipsilateral columnar projections to frontal association cortex in primates. *Science* 216:755–757.
Levitt, P., Rakic, P., and Goldman-Rakic, P. S. 1981. Region-specific catecholamine innervation of primate cerebral cortex. *Neurosci. Abs.* 7:801.
Mares, V., and Lodin, Z. 1970. The cellular kinetics of the developing mouse cerebellum. II. The function of the external granular layer in the process of gyrification. *Brain Res.* 23:343–352.
Ogren, M. P., Rakic, P., and Goldman-Rakic, P. S. 1983. Consequences of prenatal striate cortex lesions on retinogeniculate projections in the monkey. *Invest. Ophthalmol.* 23:64.
Pandya, D. N., and Sanides, F. 1973. Architectonic parcellation of the temporal operculum in rhesus monkey and its projection pattern. *Z. Anat. Entwickl.-Gesch.* 139:123–161.
Papez, J. W. 1929. *Comparative Neurology*. New York: T. Y. Crowell.
Rakic, P. 1974. Neurons in rhesus monkey visual cortex: systematic relation between time of origin and eventual disposition. *Science* 183:425–427.
Rakic, P. 1975. Timing of major ontogenetic events in the visual cortex of the rhesus monkey. In N. A. Buchwald and M. Brazier, eds., *Brain Mechanisms in Mental Retardation*. New York: Academic Press, pp. 3–40.
Rakic, P. 1976a. Differences in the time of origin and in eventual distribution of neurons in areas 17 and 18 of visual cortex in rhesus monkey. *Exp. Brain Res. Suppl.* 1:244–248.
Rakic, P. 1976b. Prenatal genesis of connections subserving ocular dominance in the rhesus monkey. *Nature* 261:467–471.
Rakic, P. 1977. Prenatal development of the visual system in the rhesus monkey. *Phil. Trans. Roy. Soc. Lond.*, Ser. B 278:245–260.
Rakic, P. 1978. Neuronal migration and contact guidance in primate telencephalon. *Postgrad. Med. J.* 54:25–40.
Rakic, P. 1979. Genesis of visual connections in the rhesus monkey. In R. Freeman, ed., *Developmental Neurobiology of Vision*. New York: Plenum Press, pp. 249–260.
Rakic, P. 1982. Early developmental events: cell lineages, acquisition of neuronal positions, and area and laminar development. *Neurosci. Res. Prog. Bull.* 20:439–451.

Rakic, P., and Goldman-Rakic, P. S. 1982. *Development and Modifiability of the Cerebral Cortex. Neurosci. Res. Prog. Bull.* 20:429-611 (edited volume for MIT Press, Cambridge, Massachusetts).

Rakic, P., and Goldman-Rakic, P. S. 1983. Use of fetal neurosurgery for experimental studies of structural and functional brain development in non-human primates. In R. T. Thompson and J. R. Green, eds., *Prenatal Neurology and Neurosurgery.* Hampton, Virginia: Spectrum Press (forthcoming).

Richman, D. P., Steward, R. M., Hutchinson, J. W., and Caviness, V. S., Jr. 1975. Mechanical model of brain convolutional development. *Science* 189:18-21.

Schmitt, F. O., ed. 1981. *The Organization of the Cerebral Cortex.* Cambridge, Massachusetts: MIT Press.

Schwartz, M. E., and Goldman-Rakic, P. S. 1982. Single cortical neurons have axon collaterals to ipsilateral and contralateral cortex in fetal and adult primates. *Nature* 299:154-155.

Sidman, R. L., and Rakic, P. 1973. Neuronal migration, with special reference to developing human brains: a review. *Brain Res.* 62:1-35.

Sidman, R. L., and Rakic, P. 1982. Development of the human central nervous system. In W. Haymaker and R. D. Adams, eds., *Cytology and Cellular Neuropathology,* 2nd ed. Springfield, Illinois: Charles C Thomas, pp. 3-145.

Sievers, J., Mangold, V., Barry, M., Allen, C., and Schlossberger, H. G. 1981. Experimental studies on cerebellar foliation. I. A qualitative morphological analysis of cerebellar fissuration defects after neonatal treatment with 6-OHFA in the rat. *J. Comp. Neurol.* 203:751-769.

Smith, G. E. 1906. New studies on the folding of the visual cortex and the significance of the occipital sulci in the human brain. *J. Anat. (Lond.)* 41:198-207.

Smith, G. E. 1931. The central nervous system. In A. Robinson, ed., *Cunningham's Manual of Practical Anatomy.* New York: Wood, pp. 505-800.

Stewart, M. R., Richman, D. P., and Caviness, V. S., Jr. 1975. Lissencephaly and pachygyria. An architectonic and topological analysis. *Acta Neuropathol.* 31:1-12.

Todd, P. H. 1982. A geometric model for the cortical folding pattern of simple folded brains. *J. Theoret. Biol.* 97:529-538.

Van der Loos, H. 1979. The development of topographical equivalences in the brain. In E. Meisami and M. A. B. Brazier, eds., *Neural Growth and Differentiation.* New York: Raven Press, pp. 331-336.

Volpe, J. J. 1981. *Neurology of the Newborn.* Philadelphia: Saunders.

Welker, W. I., and Campos, G. B. 1976. Physiological significance of sulci in somatic sensory cerebral cortex in mammals of the family Procyonidae. *J. Comp. Neurol.* 120:19-36.

Young, J. Z. 1962. *The Life of Vertebrates.* Oxford: Clardendon Press.

Part Three

Biological Associations
of Laterality

Chapter 13

Twinning, Handedness, and the Biology of Symmetry

Charles E. Boklage

The human species shares with nearly all other forms of animal life a basic body structure that is bilaterally symmetrical. In many animals the developmental elaboration of that plan includes certain specially programmed asymmetric variations. The layout of the circulatory system, the folding of the gut, and the placement of its derivative organs are prominent nonbrain examples; and in the human there is, especially, the brain. I say that these departures from symmetry are programmed because they are very clearly nonrandom — in fact, are species-specific.

Prolonged consideration of the observable asymmetries of human behavior in an effort to determine their causes leads one to contemplate a stunning array of structures, processes, relationships, implications, and conjectures. My personal reaction is to focus on a single, relatively simple aspect of the problem, and to think mechanically.

Whatever happens in the mind of man is represented in the actions and interactions of brain cells. These cells, like those of any other organ, can act only by means of a finite repertory of mechanochemical processes coded for in parts of the genome of unknown size. It seems clear that one element fundamental to the normal integrity of many human mental processes is the functional asymmetry of the brain. Where the cells of the two hemispheres act differently, there must at some level be structural differences.

It is not unreasonable to think of left and right brain hemispheres as distinctly differentiated organs, not unlike left and right chambers of the heart. For proper differentiation, which will result in the appropriate ranges of specialized function, similar but nonidentical structures have to be built in different places and connected to other structures. Each cell involved must have ways to know where it is and when the time has come to perform its appropriate functions. The distribution of this information must be highly reliable, within the individual and across generations, and must possess a measure of flexibility and mutability.

The existence of a certain level of functional plasticity adds to the complexity of the program. Some cells in the human brain can learn; some in fact must learn — that is, be programmed in part by experience. That ability, far from reducing the need to postulate an inherent developmental program, actually indicates the existence of additional layers of complexity in that program: the inherent program must include instructions for the different locations of many cells which can or must learn, and for what they are to be able to learn. At least part of the time the same cell structure can assume different functions according to its environment, but often that environment consists primarily of the structures and functions of other cells.

In order to begin to understand the biological basis of left-right brain functional asymmetry, and in particular to carve out a soluble part of that problem, let us first consider the earliest point at which we may infer irreversible establishment of the left-right body axis and then proceed to examine the effects of changes in that process.

The possible existence of special relationships between twinning and symmetry development (and thus the possibility that twinning may represent a useful anomaly by which to begin dissecting the underlying mechanisms) occurred to me via the discovery of a three-way relationship involving twinning, handedness, and schizophrenic illness. In a sample of 61 twin pairs, with at least one member of each diagnosed as schizophrenic, there were clear and statistically significant differences in several parameters of the risk, course, and content of schizophrenic illness as a function of handedness within the pairs (Boklage, 1976, 1977a). The sample was divided into (1) pairs with either or both members nonright-handed (1 − 2NRH) and (2) pairs with both members right-handed (2RH). The MZ/DZ concordance ratio was then determined (that is, the ratio of the concordance for schizophrenia in monozygotic twin pairs to the concordance in dizygotic pairs). The MZ/DZ concordance ratio was sharply elevated in the 2RH pairs (12.9) and reduced in the 1 − 2NRH pairs (0.9), relative to the ratio in the whole sample (4.2). Cases of schizophrenia in 1 − 2NRH pairs tended to be atypical and less severe by several measures. Some considered this result another strong argument against psychodynamic explanations, since removing 1 − 2NRH pairs from the sample left the result "genetic" with nearly Mendelian clarity. Others credited it with defining a new basis for a genetic subclassification.

There were also clear and significant differences between monozygotic and dizygotic pairs in handedness-schizophrenia relationships. None of the relationships between handedness and schizophrenia was statistically significant among the DZ twins (nor was there even a consistent trend), except in the case of concordance. The concordance data in DZ twins were in fact reciprocal to those among MZs. Among DZ twins 1−2NRH pairs were slightly but significantly more likely to be concordant than 2RH pairs, whereas among MZ twins 1−2NRH pairs were less likely to be concordant than 2RH pairs. 1−2NRH DZ pairs were not significantly different in concordance from 1−2NRH MZ pairs.

Profound conceptual and methodological implications stem from these zygosity differences. The genetic epidemiology of schizophrenia is in fact incomprehensible without invoking causative contributions which differ between twins and singletons or between zygosity groups of twins. (See Rao et al., 1982, for discussion of related points.)

It is probably important that all of these effects were most clearly observed on a pairwise basis, that is, individual NRH did not significantly predict any of the differences found when considering the handedness pattern of the pairs. This apparently refutes any interpretation based on "subtle brain insult" as a cause of both NRH and schizophrenia, and seems also to exclude effects of postnatal environment.

It has been argued that several relationships between brain laterality and both schizophrenia and affective psychoses convey important, but not yet fully understood, messages about their underlying biology (Gruzelier and Flor-Henry, 1979). My findings that handedness-schizophrenia relationships differ as a function of twin zygosity imply that the underlying biology of schizophrenia, or its expression in disease, also differs as a function of twin zygosity (Boklage 1977b; Elston and Boklage, 1978; Boklage et al., 1980a,b). It is not difficult to appreciate that this view implies a rejection of the fundamental assumptions of genetic twin study methodology as applied to schizophrenia. If a genetic study based on concordance in twins is to be taken at face value, and its results extrapolated to the general (primarily singleton) population, it must be assumed at the very least that the trait in question is the same trait in DZ, MZ, and singleton groups (Elston and Boklage, 1978). Since schizophrenia behaves differently in MZ and DZ groups with respect to relationships between handedness and the risk, course, and content of psychotic illness, it is hard to be comfortable with the assumption that the trait in either twin group, let alone both, correctly represents the trait in the singleton population.

It has been repeatedly asserted that NRH is found more frequently among twins than singletons, and among MZ more often than DZ twins. It has often been stated that the twin-singleton difference is due to the birth stress associated with twinning, and the zygosity difference is due to the mirror-imaging associated with some sort of splitting process believed to initiate only MZ twinning events. Birth stress, mirror-imaging, and splitting have been discussed in Boklage, 1980 and 1981a,b.

McManus (1980), upon reviewing and reanalyzing the relevant published data, concluded that there has been no sound demonstration of either twin-singleton or MZ-DZ difference of any significance. He argued that other conclusions in previous reports were the result of statistically incorrect pooling of significantly different results from different studies. His view was that only studies prior to 1930, when mirror-imaging (that is, discordance) in handedness was actually considered to be almost diagnostic of monozygosity — and zygosity diagnosis in general was rather crude — show an MZ/DZ excess of NRH at significant levels. No study since then (including Boklage, 1977a) reports a significant difference in probability of NRH between normal MZ and DZ twins without including numbers from those early studies. Excluding them, the remaining studies are equally divided on the direction of nonsignificant MZ-DZ differences.

It has also been argued that the reported twin-singleton differences derive from studies comparing data collected separately for twins and for singletons, at different times and places by different investigators with differing criteria. On the few occasions when the comparison is said to have been made directly, differences are not reported.

I have published (Boklage, 1981b) a three-generation twin-family study, involving handedness data in over 10,000 people, avoiding at least most of the problems of previous studies and allowing for some limited deliberate variation of criteria. Twins were *not* significantly more often NRH than *their own* singleton sibs, nor were frequencies in MZ and DZ groups different. To summarize the more salient results:

(1) The handedness phenotype of any individual is significantly related to that of his parents, with either parent's being NRH increasing the offspring's probability of NRH about 1.5-fold.

(2) The parents of twins are almost twice as likely as their own same-sex nontwinbearing siblings (aunts and uncles of twins) to be NRH. Taken with (1), this suggests that singletons from all-singleton families may be different, with a substantially lower probability

of NRH than singletons from families with twins. In this study, for example, among second-degree relatives of twins the probability of NRH was equal to that of general-population singletons from other recent studies. By contrast, all classes of first-degree relatives of twins had higher frequencies of NRH closer to those among twins.

(3) The secondborn member of a discordant (RH,NRH) pair of twins (whether MZ or DZ) has almost double the firstborn's probability of being the only NRH in the pair. This effect is sufficient to account for the small excess of NRH (15%, not statistically significant) among twins relative to their singleton sibs.

(4) All of the above results apply equally to both zygosity groups.

(5) MZ pairs match in handedness phenotype significantly more often than DZ pairs.

The results of this study suggest the existence of a special relationship between NRH and twinning, most (but not all) of which is heritable and has nothing to do with birth stress or any other aspect of twinship itself. It affects twins of both zygosities equally, and it is transmitted through either parent equally.

It seems, therefore, that we must deal with at least three variables: (1) the sources of a zygosity-dependent pairwise effect (point 5 above); (2) a heritable association of NRH with twinning, independent of zygosity; and (3) a minor contribution from a form of birth stress peculiar to twins born second, in either zygosity. Derom and Thiery (1976) identified the latter as a consequence of transient reduced perfusion of the second twin during delivery. That this is the basis of the within-pair birth-order effect is by no means proven, but it is plausible — and consistent with earlier, unproven suggestions in the literature.

My earlier theoretical considerations were based on results from schizophrenia in twins, with handedness effects being significant almost exclusively among MZs. These findings, taken together with the commonly held belief at that time in a higher rate of NRH among MZ than DZ twins, made it seem reasonable that some correlate of the MZ twinning process might predispose to both NRH and schizophrenia. It also seemed reasonable to suppose that MZ twinning, and not DZ, might have something to do with early embryonic determination of body and brain symmetry. In other words, something about the process of segregating cells of the early embryo into two separate developmental schemata might somehow be slightly disruptive of the symmetry-determining processes going on at the same time. These concepts provoked an attempt to demonstrate the probable time sequence of MZ twinning events (Boklage, 1981a).

The chorion and amnion are two differentiation events of approximately known timing, separated by a few days in the first week of development. When MZ twins share a particular structure, it may be assumed that that structure differentiated before the twinning event. If we accept that assumption, then the data indicate that most MZ twinning events occur between choriogenesis (about day 4) and amniogenesis (about day 7), after which a successful twinning event, with fully separate body symmetries, rapidly becomes unlikely. In this same period between days 4 and 7 the spherical inner cell mass rearranges itself to form the bilaminar disc-stage embryo. About 6 more days elapse before two features appear on the bilaminar disc which make the irreversible achievement of basic bilateral symmetry microscopically visible, and which indicate readiness to proceed toward establishment of the third embryonic cell layer. The differentiation of the prochordal plate and the primitive streak on about day 13 mark the anterior and posterior poles of the embryonic disc. The dorsoventral axis is also fixed, since the polar and mural components of the trophoblast become associated invariably with dorsal and ventral aspects respectively of the inner cell mass at the time of choriogenesis (about day 4). Fixing the anterior-posterior direction of the second of the body's three axes necessarily fixes the third as well, with the result that left and right sides are *visibly* fixed by about 2 weeks of development. The submicroscopic (that is, biochemical and/or cytoskeletal) differentiation is almost certainly fixed earlier (7.7 days, with 95% confidence limits 7.5 to 8.3 days; Boklage, 1981a), as indicated by great difficulties in recruiting cells into two fully separate body symmetries after amniogenesis. Nearly a third of all *viable* MZ pairs of postamniogenic origin are born conjoined, not having achieved full separation of body symmetries. More than 70% of such pairs are female (Derom et al., 1980; older results reviewed in MacGillivray et al., 1975). This suggests that as many as half of the postamniogenic pairs are conjoined, but the frequency is reduced by preferential loss of affected male pairs. This would move the mean time of symmetry fixation back slightly, to about 7.4 days. Thus the MZ twinning event and the initial cellular determination of body symmetry axes are quite intimately related in time, and perhaps in function as well.

Even at the time when it was accepted that MZs were more often NRH than DZs, closer consideration suggested that any such differences were small relative to the large twin-singleton differences in NRH, and smaller still relative to the differences in handedness-schizophrenia relationships between MZ and DZ twins. The subsequent realization that no MZ-DZ differences in probability of NRH can in fact be demonstrated led to an attempt to define a plausible mechanism whereby DZ, as well as MZ, twinning processes might be related to anomalies of embryonic symmetry development. Under prevailing but perhaps weakening theory, the development of DZ twins is supposed to be exactly equivalent to that of singletons, except perhaps late in pregnancy when uterine resources may be challenged. (This development is, of course, usually presumed to be exactly equivalent to that of MZ twins, beyond the very early stage in which MZ embryos are reorganized to make them into two bodies instead of one.) The "common knowledge" that DZ twins arise from double ovulation has precluded consideration of the fact that there is no compelling evidence that this is the case for all or even for the majority of DZ twins. That DZ twinning *can* result from such a process is evidenced by the rare occurrence of bilateral tubal pregnancy, which would be difficult to explain by any other mechanism. Beyond that, the assertion has survived from about the time of Aristotle by extrapolation from litter-bearing mammals. Now, however, there are several reasons for challenging this belief.

First, there is the MZ-DZ equivalence in relation to handedness in normal twins, and MZ-DZ reciprocality in some pairwise relationships between handedness and schizophrenia. The normal-twins relationship appears to be heritable; the schizophrenic-twins relationship, by its pairwise character and zygosity dependence, requires the presence of at least a prenatal and probably an early embryonic effect. The need therefore arises to ascertain possible relationships between the cellular process of embryonic symmetry determination and DZ, as well as MZ, twinning. Double ovulation, which implies that DZ twins do not differ from singletons in embryogenesis, seems an unlikely prospect.

A further problem is posed by some results of Mai (1974) on the Down syndrome in twins: "The null hypothesis that dizygous twinning and concordance for Down syndrome are independent events is tested after removal of effects such as covariance due to maternal age. The null hypothesis is rejected ($p < 0.000002$)." In other words, DZ twins both suffer from the Down syndrome significantly more often than the assumption that DZ twins have the same genetic endowment as full sibs should lead one to expect. Twinning, of both zygosities, is associated significantly with chromosomal anomalies, for example, with XO (Nance and Uchida, 1964) and with XXY (Nielsen, 1966). Chromosome segregation is mediated by the spindle generated from the centrioles and microtubule organizing centers, the same organelles responsible for determining the axes and symmetry of cell division. The axes and symmetry of cell division would have to be modified in a change from the usual asymmetry of meiotic cell divisions to the more symmetrical division probably required for polar body twinning (about which more below).

Excess DZ twinning is associated with illegitimacy, with conception very shortly after marriage, and with conception soon after return of the husband from armed service (Allen, 1981). Marriages characterized by frequent long absences, such as those of commercial fishermen and merchant mariners share this association (Eriksson, 1973). Major associations with increases in parity and maternal age have also been noted.

Allen (1981) shows that the traditional interpretation of these observations in terms of unusually high fertility or fecundity is not well founded, and suggests some psychological effect on hormone balance as their common characteristic. James (1972) postulates elevated frequency of intercourse (based on the assumption that double ovulation is rather common but that probability of double fertilization is limiting). Harlap's (1980) results imply that the important element may be a delay preceding fertile intercourse. When Orthodox Jewish Israeli women, observing the ritual abstinence from intercourse during menstruation, prolong that abstinence (which is terminated by a ritual bath) until after the expected time of ovulation, the frequency of twinning (primarily DZ) is dramatically elevated. Elevated frequencies of congenital anomalies and distortion of sex ratios at birth suggest a further association with embryonic disturbances. The sharp peak of twinning rates occurring 8 to 9 months after the peak demobilization of World War II troops was accompanied by a decided change in the sex ratio of births (MacMahon and Pugh, 1954).

Experimentally delayed fertilization in the rabbit resulted in frequent twinning and embryonic anomalies. Bomsel-Helmreich and Papiernik-Berkhauer (1976) related the observed effect to MZ twinning exclusively, but no direct demonstration of zygosity was in fact made or attempted. (It was assumed that paired blastocysts inside a single zona could only be MZ, but the same configuration would be expected from polar body DZ twinning, to be discussed below). Butcher et al. (1969) had similar results with rats.

I have previously suggested (Boklage 1978, 1980, 1981b; Elston and Boklage, 1978; Boklage et al., 1979, 1980) that polar body twinning deserves active consideration because it represents an alternative mechanism for DZ twinning that might be compatible with all of the above observations.

The specialized meiotic cell divisions which produce, in turn, the first polar body and fertilizable secondary oocyte, then the second polar body and the zygote, are extremely asymmetrical. The spindle, in the course of moving one of its poles to the extreme periphery of the cell, rotates through 90 degrees. In the ensuing division the zygote usually receives more than 95% of the cytoplasm, according to my calculations from relative diameters in photographs. The much smaller polar bodies are extruded.

Some modification of this mechanism of asymmetrical cell division is as plausible a cause of twinning as the commonly assumed double ovulation resulting from some disturbance of a complex hormonal gating mechanism, and could just as easily be heritable. It also has the advantage of being similar to mechanisms that affect patterns of cellular symmetry in development.

If the first meiotic spindle did not move to the periphery, the subsequent division would probably result in two cells of more or less equal size, genetically equivalent to secondary oocytes, in a single zona pellucida. These could be separately fertilized and then proceed normally into separate second meiotic divisions. (It obviously would be helpful if the primary oocyte were larger than usual, but we have no reason to assume that to be necessary.) By definition, this would be *first polar body twinning*. The process raises no problem with respect to the *fertilization response*, that is, the mechanism which normally blocks fertilization of the ovum by more than one sperm. Since the cells would be separated prior to fertilization, their respective fertilization responses would be distinct.

In most mammals the first meiotic division occurs just prior to ovulation, resulting in release of the secondary oocyte, which is then fertilized more or less immediately in the ampulla of the fallopian tube. The mechanism found in the rabbit, whereby ovulation is triggered by copulation, seems to be very useful, as it increases the likelihood of sperm and oocyte arriving in the ampulla together. The importance of this becomes clear when one realizes that in all species examined oocytes age rapidly after achieving readiness for ovulation and cannot usually wait more than about 12 hours if fertilization and development are to proceed normally. The timing mechanism of the rabbit seems so useful, that it is unlikely that other mammals would not share some version of this trait — perhaps less stringent and thus unnoticed (Austin, 1961).

The spontaneous resumption of the second meiotic division (normally suspended at metaphase pending fertilization) is known to occur in some species in response to aging in the unfertilized state, or upon receipt of any of several artificial stimuli. The cellular organization required to produce asymmetric cell division by way of peripheral migration of the spindle might well be destabilized by aging of the oocyte. *Second polar body twinning* could result if fertilization occurred after (rather than before) a second meiotic division that happened to be symmetrical. Here again the fertilization response, blocking polyspermy, could occur independently in the separate resulting zygotes.

Alternatively, if the fertilization response were destabilized by aging, double fertilization might occur, with pronuclei from two sperms fertilizing both secondary oocyte nuclei. This somewhat more complicated form of second polar body twinning could explain the existence of naturally occurring chimeric individuals (Dunsford et al., 1953; Nicholas et al., 1957) more simply than does fusion of DZ twin embryos. Individuals mosaic for a chromosomal anomaly are much more common than chimeras; they could arise by this dispermic second polar body mechanism as easily as by mitotic nondisjunction.

Goldgar and Kimberling (1981) elaborating a theme set forth in Elston and Boklage (1978), discussed further the difficulty involved in the genetic demonstration of polar body twinning. This would require large numbers of nonidentical twin pairs with known genotypes for a number of genes the map positions of which, at least relative to their respective centromeres, must be known. The number, map distribution, and allele frequencies of the marker genes would have strong effects on the power of the analysis. Matching genotype information from both parents would be extremely helpful. Obtaining all the requisite information would be an enormous undertaking. We require, therefore, another approach—to answer, if not all the same questions, at least some fundamental inquiries: Are MZ, DZ, and singleton developmental protocols equivalent? Or are they detectably different in ways related to symmetry development in the head?

As one possibility, we have studied a set of 56 measures from the adult dentition of MZ and DZ twins and singletons. The measurements were provided by Potter, who had previously analyzed some twin data (Potter and Nance, 1976; Potter et al., 1976). Why teeth? Because at the time when I believe embryogenic symmetries of body and brain are being fixed, the primordia of teeth, face, and telencephalon (cerebrum) are in the same cells. If twinning, and/or zygosity among twins, were either cause or consequence of differences in the processes of cellular fixation of embryonic symmetry axes, one might find demonstrable differences among these groups in developmental relationships among the teeth. It is not an ideal experiment; there are too many reasons why a difference might not be detectable in the adult.

Our methods are those of multivariate and multiple-univariate analysis. Under multiple-univariate analysis, we consider each univariate group comparison as one of multiple tests; if the groups are in fact equal, the multiple test results should have a characteristic distribution, departures from which may be tested for significance. Even when every single univariate comparison is itself nonsignificant, there may be a significant tendency for the results to be in one direction or the other. We have tested means, variances, and bivariate correlations in the original variables, and means and variances of differences in left-right pairs of these variables. In the original variables in both sexes differences among MZ, DZ, and singleton groups depart significantly from expectations of equality for one or more of the measures. In the left-right comparison, twins of both zygosities in both sexes are significantly different from singletons, and differences between MZ and DZ twins are not significant.

Under multivariate methods, the entire set of observation matrices, involving simultaneously all of the variables and all of the linear relationships among them, can be considered as integrated patterns and examined for differences as a function of group membership. Linear discriminant function analysis focuses on equality of multivariate group mean vectors, under the assumption of equal group covariance matrices. Quadratic discriminant function analysis tests equality of covariance matrices as well. The sphericity test deals with the possibility that covariance matrices may be equal except for a constant of proportionality. Discriminant analysis also produces a classification function, constructed to maximize separation between the input reference groups. With this function all observations, regardless of whether they are part of an input reference group, are classified according to probable membership in such a group. According to the results of these tests, MZ-DZ-singleton group differences are of sufficient strength and consistency to allow placement of an individual in one of 6 sex-twinship-zygosity classifications with very nearly 100% accuracy. These analyses need repetition with additional samples, our realistic expectation being that the ultimate classification accuracy achieved may be slightly less than that obtained with the samples.

The statistical significances of the various tests of multivariate distributional equality leave little doubt that MZ $\neq$ DZ $\neq$ singleton in overall craniofacial development as measured from structural relationships in secondary dentition. For example, zygosity can be determined for an individual twin entirely without cotwin comparison information, as accurately as by genotyping and comparing some 20 markers on both twins. The differences between the singleton and DZ groups are in general larger than the corresponding MZ-DZ differences, but are not less than the differences between singletons and MZs. In other words, DZs are no more like singletons than MZs are, with respect to overall protocols of individual craniofacial development. There are noticeable size differences in males and females, but other results are much less marked; one can, with practice, look at a set of these measurements and make a good guess about gender, but not about twinship or zygosity. These results are detailed in Boklage (1983a).

Where does symmetry development fit into this picture? These classification functions are weighted sums of the input variables, each variable multiplied by a loading coefficient. The relative loadings of antimeric pairs of variables in the discriminating equations computed in these analyses tend strongly to be of opposite sign, making it appear that symmetry is a factor in some of the discriminations. In further analyses focused more directly on left-right differences within individual sets of measurements, we find that twins, to similar extents in both zygosity groups, are individually more symmetrical than singletons, so that there are substantial reductions in both directional and fluctuating components of asymmetry variance. The directional component reflects the nonzero mean left-right difference; the fluctuating component reflects individual variation about the group mean left-right difference. Significant tendencies for one side to be larger that are observed in singletons are often reduced, sometimes even reversed, in the dental diameters of the twins in whom left-right pairs of variables tend significantly to be more alike and more predictably related (Boklage, 1983b). Taken together with the handedness study (Boklage, 1981b), we have two results, in different samples, with different questions (behavioral in the one case and structural in the other), both indicating a reduction from the usual degree of asymmetry in twins of both zygosities relative to singletons.

Overall these results seem to require a conclusion that twinning is indeed associated with unusual distributions of symmetry development in normal and abnormal brain function, and in craniofacial anatomy. They do not yet tell us much about detailed mechanisms, but rough outlines seem to be emerging.

If these results applied only to MZ twins, the interpretation would be substantially easier. MZ twinning, more or less obviously, is a duplication anomaly of body symmetry fixation: two ova are fertilized at the same time — and two plans are laid out — instead of one, in a way apparently analogous to duplication of any other differentiating body structure component. When we have been able to examine some of these questions directly, we have concluded that DZs are approximately as different from singletons as MZs are. Some might argue that the differences between twins and singletons result not from different paths through embryogenesis, but from some feature of twin gestation itself. This view is severely weakened by the realization that most of the anomalous sequelae of twin pregnancy (such as increases in malformations and infant mortality) differ dramatically with sex and zygosity (Hay and Wehrung, 1970; Myrianthopoulos, 1976; Hyde et al., 1980), whereas the effects we observe with respect to symmetry development are little influenced by those factors.

All these results have led me to look carefully at embryogenesis. Our results are simply not compatible with the hypothesis that double ovulation is the primary source of DZ twinning and that it is followed by normal embryogenesis. An explanation based on polar body twinning offers an alternative, according to which both DZ and MZ twinning are anomalies of the processes that determine symmetry (Boklage, 1978, 1980). In most organisms examined to date, little or no function of the zygote genome (as measured by expression of paternal genes or synthesis of new messenger RNA) can be detected prior to gastrulation (Davidson, 1976), that is, before about day 14 in the human. The results described above place all twinning events and all of the events of symmetry fixation before day 14 — which would imply that those events might be determined by structures and information already present in the gametes, placed there under the control of the parental genomes.

I have inferred that body symmetry is usually fixed during the eighth day of embryonic development. This is a moderately reliable process, carried out by cells that cannot reliably do anything they do not already "know" how to do. Part of the repertory is in the genome, part in gametic templates built by the parental genomes, in templates built by the grandparental genomes, and so on. Twinning, arising from subtle perturbations of the processes that determine symmetry, may narrow our search for those processes. The changes that produce the differences discussed here have to be small, to result in a plan which is subtly different from the usual but still works. The corresponding changes at the cytoskeletal and mechanochemical level will not be easy to find, but I am convinced that many of the abiding questions in human mental and behavioral development cannot be resolved until some of these more basic questions are answered.

Support for the work described here has been provided by the Biology Division of Kansas State University, the Genetics Curriculum of the University of North Carolina School of Medicine, and the dean of the East Carolina University School of Medicine. The eager and competent assistance of the National Organization of Mothers of Twins Clubs in harvesting data from over 800 three-generation twin families is gratefully acknowledged. Coding and computational assistance from JoAnne Mills, Des Laux, Joy Knox, and Bill Baker has been crucial to progress.

References

Allen, G. 1981. The twinning and fertility paradox. In L. Gedda, P. Parisi, and W. E. Nance, eds., *Twin Research 3, Part A: Twin Biology and Multiple Pregnancy*, vol. 69A of *Progress in Clinical and Biological Research*. New York: Alan R. Liss, pp. 1–13.

Austin, C. R. 1961. *The Mammalian Egg*. Oxford: Blackwell Scientific Publications.

Boklage, C. E. 1976. Embryonic determination of brain programming asymmetry: a neglected element in twin-study genetics of human mental development. *Acta Genet. Med. Gemellol.* 25:244–248.

Boklage, C. E. 1977a. Schizophrenia, brain asymmetry development, and twinning: cellular relationship with etiological and possibly prognostic implications. *Biol. Psychiatry* 12:19–35.

Boklage, C. E. 1977b. Embryonic determination of brain programming asymmetry—a caution concerning the use of data on twins in genetic inferences about mental development. *Ann. N.Y. Acad. Sci.* 299:306–308.

Boklage, C. E. 1978. On cellular mechanisms for heritably transmitting structural information. *Behav. Brain Sci.* 2:282–286.

Boklage, C. E. 1980. The sinistral blastocyst: an embryologic perspective on the development of brain-function asymmetries. In J. Herron, ed., *Neuropsychology of Lefthandedness*. New York: Academic Press, chap. 3.

Boklage, C. E. 1981a. On the timing of monozygotic twinning events. In L. Gedda, P. Parisi, and W. E. Nance, eds., *Twin Research 3, Part A: Twin Biology and Multiple Pregnancy*, vol. 69A of *Progress in Clinical and Biological Research*. New York: Alan R. Liss, pp. 155–165.

Boklage, C. E. 1981b. On the distribution of nonrighthandedness among twins and their families. *Acta Genet. Med. Gemellol.* 30:167–187.

Boklage, C. E. 1983a. Differences in protocols of craniofacial development due to twinship and zygosity among twins. Submitted to *J. Craniofacial Genet. Devel. Biol.*

Boklage, C. E. 1983b. Effects of twinship and zygosity on directional and fluctuating asymmetry in craniofacial development. In preparation.

Boklage, C. E., Elston, R. C., and Potter, R. H. 1979. Cellular origins of functional asymmetries: evidence from schizophrenia, handedness, fetal membranes, and teeth in twins. In J. H. Gruzelier and P. Flor-Henry, eds., *Hemisphere Asymmetries of Function in Psychopathology.* London: Elsevier/North Holland, pp. 79–104.

Boklage, C. E., Elston, R. C., and Potter, R. H. 1980a. Zygosity-related differences in developmental integration. Third International Congress on Twin Studies, Jerusalem. Abstract in *Acta Genet. Med. Gemellol.* 29:71.

Boklage, C. E., Elston, R. C., and Potter, R. H. 1980b. Methodological implications of zygosity differences in developmental relationships. Third International Congress on Twin Studies, Jerusalem. Abstract in *Acta Genet. Med. Gemellol.* 29:28.

Bomsel-Helmreich, O., and Papiernik-Berkhauer, E. 1976. Delayed ovulation and monozygotic twinning. *Acta Genet. Med. Gemellol.* 25:73–76.

Butcher, R. L., Blue, J. D., and Fugo, N. W. 1969. Overripeness and the mammalian ova. III. Fetal development at midgestation and at term. *Fertil. Steril.* 20:222–229.

Davidson, E. H. 1976. *Gene Activity in Early Development.* 2nd ed. New York: Academic Press.

Derom, R., and Thiery, M. 1976. Intrauterine hypoxia — a phenomenon peculiar to the second twin. *Acta Genet. Med. Gemellol.* 25:314–316.

Derom, R., Thiery, M., and Vlietinck, R. 1980. Sex ratio of twins according to zygosity and placental structure. Third International Congress on Twin Studies, Jerusalem. Abstract in *Acta Genet. Med. Gemellol.* 29:40.

Dunsford, I., Bowley, C. C., Hutchinson, A. M., Thompson, J. S., Sanger, R., and Race, R. R. 1953. A human blood-group chimaera. *Brit. Med. J.* 2:81.

Elston, R. C., and Boklage, C. E. 1978. An examination of fundamental assumptions of the twin method. In W. E. Nance, ed., *Twin Research, Part A: Psychology and Methodology,* vol. 24A of *Progress in Clinical and Biological Research,* New York: Alan R. Liss.

Eriksson, A. W. 1973. Human twinning in and around the Åland Islands. *Commentationes Biologicae* 64. Helsinki: Societas Scientiarum Fennica.

Goldgar, D. E., and Kimberling, W. J. 1981. Genetic expectations of polar body twinning. *Acta Genet. Med. Gemellol.* 30:257–266.

Gruzelier, J. H., and Flor-Henry, P., eds. 1979. *Hemisphere Asymmetries of Function in Psychopathology.* London: Elsevier/North Holland.

Harlap, S. 1980. Twin pregnancies following conceptions on different days of the menstrual cycle. Third International Congress on Twin Studies, Jerusalem. Abstract in *Acta Genet. Med. Gemellol.* 29:40.
Hay, S., and Wehrung, D. A. 1970. Congenital malformations in twins. *Am. J. Hum. Genet.* 22:662.
James, W. H. 1972. Coital rates and dizygotic twinning. *J. Biosoc. Sci.* 4:101–105.
Layde, P. M., Erickson, J. D., Falek, A., and McCarthy, B. J. 1980. Congenital malformations in twins. *Am. J. Hum. Genet.* 32:69–78.
MacGillivray, I., Nylander, P. P. S., and Corney, G. 1975. *Human Multiple Reproduction.* London: W. B. Saunders.
MacMahon, B., and Pugh, T. F. 1954. Sex ratio of white births in the United States during the Second World War. *Am. J. Hum. Genet.* 6:284–292.
McManus, J. C. 1980. Handedness in twins: a critical review. *Neuropsychologia* 18:347–355.
Mai, L. 1974. Down's syndrome in twins: lack of evidence for independence of nondisjunction and dizygotic twinning. *Anthropology UCLA* 6.
Myrianthopoulos, N. C. 1976. Congenital malformations in twins. *Acta Genet. Med. Gemellol.* 25:331–335.
Nance, W. E., and Uchida, I. 1964. Turner's syndrome, twinning, and an unusual variant of glucose-6-phosphate dehydrogenase. *Am. J. Hum. Genet.* 16:380–390.
Nicholas J. W., Jenkins, W. J., and Marsh, W. L. 1957. Human blood chimeras: a study of surviving twins. *Brit. Med. J.:* 1:1458.
Nielsen, J. 1966. Twins in sibships with Klinefelter's syndrome. *J. Med. Genet.* 3:114–116.
Potter, R. H., and Nance, W. E. 1976. A twin study of dental dimension. I. Discordance, asymmetry, and mirror imagery. *Am. J. Phys. Anthrop.* 44:391–395.
Potter, R. H., Nance, W. E., Pao-Lo Yu, and Davis, W. B. 1976. A twin study of dental dimension. II. Independent genetic determinants. *Am. J. Phys. Anthrop.* 44:397–412.
Rao, D. C., McGue, M., and Gottesmann, I. I. 1982. Resolution of the genetic and cultural transmission of schizophrenia based upon the analysis of European family studies. 33rd Annual Meeting, American Society of Human Genetics. Abstract 104, *Am. J. Hum. Genet.* 34:44A.

Chapter 14

Laterality, Hormones, and Immunity

Norman Geschwind
Peter O. Behan

Looking back over the history of neurology and medicine since their establishment in the middle of the last century, one finds a group of research fields that developed independently. As we have seen, the study of cerebral dominance grew out of the discovery of Paul Broca that aphasia-producing lesions were found predominantly on the left side of the brain. The beginnings of neuroimmunology also date back to the last half of the nineteenth century.

One of Louis Pasteur's many accomplishments was the development of a vaccine against rabies, a disease which today seems almost exotic to those who live in England or the United States. Yet its importance on the Continent led Pasteur to develop a vaccine against the then-undetected agent. He injected rabies-infected material intracerebrally in rabbits. After repeated passages the virus changed so that it lost its capacity to generate clinical disease but could still produce immunity to the natural, so-called street virus. The vaccine injected was extracted from rabbit spinal cords in which the virus had been grown and was apparently effective in preventing clinical rabies. Before long, however, it was discovered that some patients who received the vaccine developed "neuroparalytic accidents." The mechanism soon emerged from the work of Centanni (1898) and Aujesky (1900), who discovered that the injection of brain material into animals often led to convulsions and paralysis, and that the recipients might become immunized against central nervous tissue. These early discoveries constituted the beginnings of the field known today as neuroimmunology.

Another field that expanded rapidly was endocrinology, a discipline which in its early years developed independently of discoveries in neurology.

The three fields—cerebral dominance, endocrinology, and immunity—continued to advance in apparent isolation for many years. Indeed, the role of brain mechanisms in endocrinology was not evident until well into the twentieth century. Similarly, the interrelationship of endocrinology and immunity first reached prominence when the adrenal steroids were introduced into medicine in the early 1950s as a treatment for severe rheumatoid arthritis. Despite the striking sex differences in the rates of many immunological disorders and in response to foreign substances, the association of sex hormones and immunity received no more than passing attention.

The study of cerebral dominance proceeded on its own course, virtually unaffected by parallel developments in the elucidation of fundamental biological mechanisms in normal function and disease. Cerebral laterality was somehow regarded as relevant only to certain aspects of the cognitive functions of the cortex.

All this has now been changed, for the three fields have been shown to have a wide variety of interrelationships. Occasional papers documented a relationship between one manifestation of dominance—handedness—and various noncognitive manifestations. Differences were reported in bodily structure or in various diseases encountered among left-handers. Many of these publications lacked adequate data; some simply reflected old prejudices against left-handers. Other articles, however, deserved much more than the neglect with which they were greeted. Tisserand (1949) called attention to the fact that harelip is found on the left side in about two-thirds of the cases seen. Furthermore, about 20% of the harelip patients were left-handed as against 8% of controls, and those harelip patients who were right-handed had a much higher proportion of first-degree left-handed relatives than did unaffected right-handers.

Other noncognitive associations were described. Many investigators reported that left-handedness was more common in males, although this was often dismissed as the result of differential susceptibility to social pressure of the two sexes. Similarly, left-handedness was reported as being more common in twins, although this was often dismissed as the result of increased likelihood of brain damage during birth. Finally, studies of fingerprints revealed differences between left-handers and right-handers.

The one association of left-handedness that was repeatedly mentioned was its linkage with the developmental disorders of childhood, such as dyslexia, stuttering, delayed speech, and childhood autism. More recently, similar correlations have been described with hyperactivity and the Tourette syndrome. Even these were apparently controversial, for many rejected the findings of series documenting elevated rates of sinistrality in these conditions. Our own data bring strong additional evidence that left-handedness is closely linked to these conditions. Even when this relationship was at last accepted, it only confirmed the belief that dominance was related exclusively to cognitive function and dysfunction.

Another group of papers reported associations between certain diseases and the childhood developmental disorders (but left-handedness was not reported as a factor). Several documented elevated rates of the childhood allergies in stuttering, of celiac disease or some similar disorder in childhood autism, and high rates of food allergy in hyperactivity. These relationships have often been regarded as causal. Thus some have argued that there is a condition of "celiac autism"; that is, it is assumed that celiac disease may have autism as one of its manifestations. Some have argued that hyperactivity is itself the result of food allergy, so that antiallergic treatment has been advocated as therapy for this condition. As we shall suggest, it is much more likely that the immune disorders do not cause the developmental cognitive disorders, but rather that both are the parallel effects of some common factor which favors the occurrence of sinistrality, learning disorders, and immune disease.

The studies to be reported grew out of the observations made by one of us (NG) subsequent to a November 1980 meeting of the Orton Dyslexia Society. NG commented at that meeting that the genetics of dyslexia should probably not be examined in isolation; that is, one should study not only the frequency of dyslexia among relatives of dyslexics, but also the presence of other conditions in these families. A large number of those attending afterward told NG about their family histories, and the apparently high rate of immune disorders and migraine in these accounts led to more detailed observations. It became apparent that the relationship was not only to the childhood developmental disorders, but more broadly to left-handedness. As a result, the formal studies reported here were initiated.

There are several points to be made before we present our data. In the first place, the elevated rates of various diseases and of learning disorders in sinistrals may lead to the conclusion that left-handers in general are less healthy or less adequate intellectually. One must, however, be aware that a group with an elevated rate of certain conditions may also have low rates of other disorders. Morbidity and death from malaria, for instance, are probably less frequent in families carrying the sickle-cell gene, and this probably more than counteracts the elevated morbidity and death rates of those with sickle-cell anemia. Many diseases are found only in women, yet females have lower overall morbidity and mortality at all ages than males because they have lower rates of such disorders as myocardial infarction and cancer of the lung.

It is our belief that left-handers, despite their elevated rates of certain disorders, probably have lower rates of other conditions than right-handers. Furthermore, despite the elevated rate of the childhood developmental disorders, sinistrals are found in much higher numbers than right-handers in certainly highly skilled occupations (architecture, for example). Thus they probably do not suffer an overall disadvantage in either health or cognitive function.

An additional point deserves emphasis. It is easy to be misled by the fact that the studies reported here have compared strongly left-handed and strongly right-handed people, the strong sinistrals being found to have a much higher rate of immune disease and learning disability than the dextrals. We have also found a higher rate of these disorders in the first-and second-degree relatives of the sinistrals than in those of the dextrals. Many of these affected relatives are, however, right-handed. The reason is that left-handedness is probably only one marker of *anomalous dominance*. While space does not permit a full discussion, it seems likely that about 70% of the population has standard brain dominance, that is, strong left-hemisphere dominance for language and handedness. By contrast, about 30% of the population probably has anomalous dominance, that is, some deviation from this standard pattern (more nearly equal language abilities in both hemispheres, or better language capacity in the right hemisphere, or manual skills other than strong right-handedness). The frankly left-handed probably constitute only about a third of those with anomalous dominance. Thus some of those with anomalous dominance will certainly be right-handed in the face of an altered pattern of language dominance. Why then study left-handedness? The answer is that it is still the most easily studied marker for anomalous dominance. Probably most, if not all, individuals with developmental learning disorders have anomalous dominance, even though most of them are not frankly sinistral.

Handedness is a useful but potentially treacherous marker, inasmuch as many of those with anomalous language dominance are right-handed. For this reason we compared in our study only those who on the Oldfield Handedness Inventory scored either -100 (all responses were left-handed) or $+100$ (all responses were right-handed). This provided us with much more homogeneous groups. Our left-handers with scores of −100 all have anomalous dominance. Our +100 right-handed groups may still contain some individuals with anomalous language dominance, but probably not very many.

We conducted two studies in Glasgow. The first of these, carried out in 1981 and 1982, has already been reported (Geschwind and Behan, 1982). It comprised two separate investigations in which a total of 500 strong sinistrals and 900 strong dextrals were compared. (Table 14.1 summarizes the data of the two parts of this study.) In the first part of this study we compared the responses of strongly sinistral individuals patronizing a shop in London selling instruments for left-handers with strongly dextral individuals collected in the general population of Glasgow. In this study the strong left-handers had about 2.5 times as high a rate of immune disorders and about 10 times as high a rate of learning disabilities as the strong right-handers; in addition, both immune disorders and learning disabilities were more common in the relatives of sinistrals than in those of the dextrals.

Table 14.1 Occurrence of immune and learning disorders in strong left-handers and right-handers.

	Those with immune disorders			Those with learning disabilities		
	No. left-handers (%)	No. right-handers (%)	p	No. left-handers (%)	No. right-handers (%)	p
Part 1 of study:[a]						
Subjects	27 (10.7)	10 (4.0)	<.005	24 (9.5)	2 (0.8)	<.005
First-degree relatives	48	25	<.01	21	7	<.01
Second-degree relatives	45	23	<.01	11	1	<.001
Part 2 of study:[b]						
Subjects	13 (5.3)	15 (2.3)	<.025	27 (10.9)	8 (1.2)	<.001
First-degree relatives	58	102	<.025	19	24	<.025
Second-degree relatives	19	18	<.005	2	4	c

[a] Left-handers: N = 253; right-handers: N = 253.
[b] Left-handers: N = 247; right-handers: N = 647.
[c] Numbers too small for calculation of p.

In the second part of the first study we collected both the sinistrals and the dextrals from the general population in Glasgow. We accepted the diagnosis of immune disorder only if it had been made in the teaching hospitals in Glasgow. This, of course, reduced our absolute numbers, since patients with such conditions as rheumatoid arthritis or autoimmune hypothyroidism might have been seen only by the general practitioner. The ratios, however, remained unchanged; immune disorders were still about 2.5 times as frequent in the strong left-handers.

Let us look now at what our second study revealed.

Subjects

Of the total of 1,396 subjects who formed the basis of this study, there were 652 strongly right-handed controls. These comprised 278 males and 374 females, ages 12 to 50 years, who had Oldfield laterality quotients (LQ) of +100 (fully right-handed). There were 440 strongly left-handed subjects—180 males and 260 females, ages 11 to 50 years, with LQ of −100 (fully left-handed). In addition, we studied 304 patients with proven autoimmune diseases (130 males and 174 females, ages 16 to 54 years) who were being followed in hospital clinics. All the subjects were from Scotland, except for 280 who visited a shop for left-handed people in London. An additional 1,000 randomly selected individuals completed only the handedness inventory. The second study was carried out in 1982 and 1983.

Methods

A single investigator was responsible for completion of the questionnaires. Answers to a series of questions about personal and familial medical history, together with the Oldfield Handedness Inventory (Oldfield, 1971), were filled in by each subject tested. The score on the inventory was expressed as a laterality quotient, ranging from +100 (complete right-handedness in all tasks) to −100 (complete left-handedness in all tasks). The chi-square test was used for statistical comparisons.

Results

As can be seen in Table 14.2, there were 440 left-handers (LQ = − 100) and 652 right-handers (LQ = + 100). Migraine, allergies, dyslexia, stuttering, skeletal malformations, and thyroid dis-

Table 14.2 Comparison of strong left-handers and right-handers.

Disorder	Left-handers (N = 440) No.	%	Right-handers (N = 652) No.	%	p	Ratio of rates (left-handers to right-handers)
Migraine	66	15	46	7	< 0.001	2.1
Allergies	31	7	4	0.6	< 0.001	11.5
Dyslexia	31	7	2	0.3	< 0.001	23.3
Stuttering	20	4.5	6	0.9	< 0.001	4.9
Skeletal malformations	9	2	4	0.6	< 0.005	3.3
Thyroid disorders	13	3	7	1	< 0.02	2.8
Depression	18	4	20	3	N.S.	1.3
Premature graying	18	4	20	3	N.S.	1.3

orders were increased to a statistically significant degree in the left-handed group. Apart from the dyslexic group, in which males predominated, females predominated in the other subgroups: in migraine the female-to-male ratio was 4:1; thyroid disorders, 4:1; allergies, 2:1; skeletal malformations, 3:1. If the dyslexics and stutterers are pooled, there is a rate of 11.6% in the left-handed group and 1.2% in the right-handed group, figures closely comparable to those for learning disabilities in the two groups in both of the earlier studies.

In both parts of the original study and in the second study we found a higher rate of autoimmune diseases in the left-handers than in the right-handers. Certain diseases — those involving the gastrointestinal tract and the thyroid — appeared to be especially frequent in the left-handers. In order to test this impression we studied the frequency of left-handedness in each of 8 groups of patients with different autoimmune diseases and compared these with the frequency found in 1,000 general population controls collected at the same time. Employing the conventional criterion for left-handedness (LQ < 0), we found a rate of 10.6% in the controls. Among the 304 patients with autoimmune diseases 5 of the groups had a rate of left-handedness significantly higher than that of the controls (Table 14.3). In conformity with our expectation, the highest rates of left-handedness were found in the gastrointestinal immune disorders (celiac disease, ulcerative colitis, regional ileitis) and those involv-

Table 14.3 Rate of left-handedness in patients with autoimmune disease (N = 304).

Disorder	No. of cases	Left-handed[a]			First-degree left-handed relatives[b] (%)
		No.	%	p	
Crohn's disease	30	10	33	<.0003	63
Celiac disease	36	11	31	<.0003	61
Thyroid disorders	40	9	23	<.01	25
Ulcerative colitis	32	7	22	<.03	38
Diabetes mellitus	20	4	20	N.S.	25
Myasthenia gravis	36	7	19	<.05	44
Rheumatoid arthritis	80	11	14	N.S.	15
Polymyositis	30	3	10	N.S.	20

a. LQ = 0 to − 100.
b. Percentage of individuals in each group with first-degree left-handed relatives.

ing the thyroid. The sex ratios were approximately equal for these illnesses, except in thyroid disorders and rheumatoid arthritis, where the female-to-male ratio was 2:1, and in myasthenia gravis, where it was 3:1.

Discussion

In all, we have now studied nearly 3,000 subjects. The latest results have confirmed the statistically significant associations between left-handedness, developmental learning disorders, migraine, and autoimmune diseases. Although in the past the correlation between sinistrality and dyslexia or stuttering has been questioned, our technique of comparing only strong left-handers and strong right-handers (thus avoiding the problems of classifying those with intermediate scores) confirms the presence of a powerful effect of sinistrality on the frequency of these disorders. An understanding of the mechanism that causes left-handedness is therefore of central importance in the study of learning disabilities.

Migraine was also present at elevated rate in the strongly left-handed subjects. This diagnosis was made only when severe, recurrent headache associated with visual disturbances, nausea, anorexia, and occasional vomiting were present, often with a family history. Common and classic varieties with sensory or motor prodromes were included. In strong right-handers the frequency of migraine was 7%, which is comparable to that reported for both sexes in the general population (Refsum, 1975). Recent theories of the pathogenesis of migraine add to the significance of these findings. A genetic element has long been postulated, as has a relationship with allergy (De Gowin, 1973). A carefully controlled study (Egger et al., 1983) of children with severe migraine showed that the headaches could be abolished if certain foods were avoided — a finding compatible with, although not diagnostic of, an allergic cause. An association between atopic disorders and stuttering has been reported (Diehl, 1958). Among the strong left-handers we also detected an elevated rate of skeletal malformations, especially those involving the midline (such as skull defects, scoliosis, and spina bifida occulta).

In another study of 130 families with one dyslexic member or more we found two siblings of dyslexics with transposition of the great vessels. In each of these cases one of the parents was left-handed. One of us has discussed this association (Behan, 1982), and further study will be needed to ascertain its relevance. Among children of women with lupus erythematosus, another autoimmune disease, there is a high rate of congenital abnormalities, including transposition of the great vessels (Esscher and Scott, 1979).

The close relationship of neurological and immune development is reflected in the many conditions in which there are abnormalities in both systems. Hagberg and colleagues (1970) reported a brother and sister with motor dysfunction and defective thymus-dependent immunity in whose brains there was neuronal dysplasia with heterotopias. Similar cases have been reported from Glasgow (Graham-Pole et al., 1975), with motor abnormalities and defective cellular immunity. These disorders show obvious similarities to ataxia-telangiectasia, the most familiar disorder in which brain and thymus are both affected; gonadal abnormalities may also be present.

Laboratory studies provide further evidence for a close relationship between nervous and lymphoid tissue. Studies with monoclonal antibodies have shown cross-reactivity between fetal brain, thymic and lymphoid tissue, and cerebral tumors (Wikstrand and Bigner, 1982). Antigens such as Thy-1 are expressed on brain and thymic cells, while antibrain antisera will cross-react with lymphoid and hemopoietic cells (Reif and Allen, 1966; Golub, 1972). It has been suggested that the microenvironment in the brain may be similar to that in the bone marrow, because pluripotential hemopoietic stem cells have been identified in the central nervous system (Bartlett, 1982). The detection of similar cells in the fetal thymus, as early as 15 days after conception, indicates that during development there may be migration from thymus to brain. It has even been suggested that a primordial cell may give rise to both lymphoid and neural elements (Bartlett, 1982). These findings make it easy to understand how a common influence may alter the brain as well as the immune system.

We have postulated that certain intrauterine influences may delay the development both of the left hemisphere and of the thymus (Geschwind and Behan, 1982). The common influences thus lead simultaneously to an increased rate of left-handedness and of susceptibility to immune disorders. When the retarding influence is even more marked, neuronal migration defects may be produced, leading to childhood dyslexia, as in the case reported by Galaburda and Kemper (1979). The evidence for the hypothesis that this retarding influence might be testosterone is analyzed in greater detail in Geschwind and Behan (1982), which also summarizes many of the other known relationships between testosterone and immunity.

There is now further evidence for such a relationship: the number and titer of autoantibodies shows an inverse relationship to serum levels of testosterone in cirrhotic male patients (Gluud et al., 1981), and there are receptor sites for sex hormones on thymic epithelium (Talal et al., 1981). In earlier studies of the effects of testosterone on development of the immune system postnatal administration of this hormone led to alterations in the thymus. Although we have not found reports of the effect of testosterone on the development of the thymus in utero, we are studying this problem.

In a preliminary investigation testosterone was administered during the first 4 days of pregnancy to inbred strains of rats and outbred rabbits, at doses ranging from 150 μg to 2 mg per 100 g body weight. Control rats and rabbits, injected only with arachis oil, delivered normally; but females rats receiving testosterone all cannibalized their litters. The same phenomenon was observed in the rabbits, but 10 litters were salvaged and raised. Serial tests for antithyroglobulin antibodies, using the ELISA technique, showed a conspicuous increase in number and titer by 6 weeks of age, compared to controls. The preliminary results are consistent with the view that testosterone in utero affects development of the thymus.

Another recent study has reported a surprising interaction between the immune system and dominance. Renoux and associates (1983) have shown that the acquisition of lymphocyte receptors (Thy-1, for instance) requires an intact *left* cerebral cortex. Ablation of the cortex on the left led to imbalance in T-cell-mediated responses. The investigators suggest that the cortex produces a factor which has a direct effect on T-cell function. Since this factor depends on the left hemisphere, the data support the belief that dominance is related to immunity.

Several other findings may be related to these novel observations. We published data some years ago showing that patients with benign or malignant brain tumors had impaired T-cell function (Thomas et al., 1975; Menzies et al., 1980), and this has been confirmed elsewhere (Brooks et al., 1972). It was difficult at the time of the original studies to understand why a benign brain tumor in an otherwise healthy individual would be associated with such immune abnormalities. Our studies and those of Renoux and coworkers (1983) suggest plausible hypotheses, testable experimentally. Intrauterine influences might affect brain and immune development in parallel fashion and thus lead to the association of brain tumors and immune anomalies. Another possibility is that local effects of the brain tumor might lead to the observed changes. Brooks and colleagues (1972) found serum factors in the patients and thought that these might be antibodies or antigen-antibody complexes. Since we have been unable to confirm this hypothesis, it is probable that a serum factor of the type postulated by Renoux may be responsible. We have also studied lymphocyte function and lymphocyte subpopulations in patients with brain damage and found evidence of impairment. The mechanisms by which T-cell changes are induced preferentially by lesions on one side require further investigation.

Our findings have now received confirmation by other investigators. Kolata (1983) has reported on the work of Camilla Benbow and Julian Stanley of Johns Hopkins University. These investigators had found that among mathematically highly gifted children there was a marked excess of males, which suggested a possible role of male hormones. They have now shown that the rate of left-handedness in this group is about twice as high and the rate of allergies about 6 times as high as those in children of average mathematical endowment. Lawrence Wood of Boston reported at the American Thyroid Association meeting in October 1983 that the rate of left-handedness was significantly higher in patients with immune thyroid disease than in those with nonimmune thyroid disorders. Marcel Kinsbourne and Brenda Bemporad (personal communication) have found a higher rate of immune disorders in relatives of dyslexic children who were left-handed than in those who were right-handed.

Some Final Comments

The existence of an association between patterns of cerebral dominance and disorders of the immune system remains as surprising to many who first hear about it as it was at the moment of its discovery. Like other observations made in the past, such as the high frequency of left-handedness in twins and in children with harelip, or the asymmetries in the brain of the trout, the cod, and the lamprey, this new finding could have rapidly slipped into oblivion. As long as fundamental biological knowledge was lacking, cerebral dominance remained an oddity, a curious exhibit of the type found in museums devoted to esoteric subspecialties.

As this volume illustrates, however, once knowledge concerning laterality becomes part of the larger field of biology, phenomena even more suprising than those neglected in the past come repeatedly to light. This chapter and the preceding ones have documented many observations, but many others have had to be omitted. What was only recently an apparently special feature of the human brain has come to be important in the study of evolution and embryology, in comparative zoology, in anatomy, pharmacology, physiology, endocrinology, and immunology — indeed, probably in every branch of biology and medicine. We even suggest that as one pushes back to the fundamental determinants of asymmetry at the molecular biological level, one will find mechanisms common to the preferential unilateral rotation of many single-celled organisms, the species-specific spiral patterns of growth of many plants, and the initial stages of formation of the asymmetrical nervous system of the human during embryonic development.

References

Aujesky, A. 1900. Ueber Immunizierung gegen Wut mit normaler Nervensubstanz. Cited in Hurst (1932).
Bartlett, P. F. 1982. Pluripotential hemopoietic stem cells in adult mouse brain. *Proc. Natl. Acad. Sci. USA* 79:2722-25.
Behan, P. O. 1982. Left-handedness. *Brit. Med. J.* 285:652.
Behan, P. O., and Behan, W. M. H. 1982. Neuroimmunology. *Practitioner* 226:2044-52.
Behan, W. M. H., Behan, P. O., and Durward, W. F. 1981. Complement studies in migraine. *Headache* 21:55-57.
Brooks, W. H., Netsky, M. G., Normansell, D. E., and Horwitz, D. A. 1972. Depressed cell-mediated immunity in patients with primary intracranial tumors. *J. Exp. Med.* 136:1631-47.
Centanni, 1898. The effects of the injection of normal brain emulsion into rabbits, with special reference to the etiology of the paralytic accidents of antirabies treatment. Cited in Hurst (1932).
De Gowin, E. L. 1973. Allergic migraine: a review of 60 cases. *J. Allergy* 3:557-566.

Diehl, C. F. 1958. *A Compendium of Research and Theory on Stuttering.* Springfield, Illinois: Charles C Thomas, pp. 83-85.
Egger, J., Wilson, J., Carter, C. M., and others. 1983. Is migraine food allergy? *Lancet* 2:865-869.
Esscher, E., and Scott, J. S. 1979. Congenital heart block and maternal systemic lupus erythematosus. *Brit. Med. J.* 1:1235-38.
Galaburda, A. M., and Kemper, T. L. 1979. Cytoarchitectonic abnormalities in developmental dyslexia: a case study. *Ann. Neurol.* 6:94-100.
Geschwind, N., and Behan, P. 1982. Left-handedness: association with immune disease, migraine, and developmental learning disorder. *Proc. Natl. Acad. Sci. USA* 79:5097-5100.
Gluud, C., Tage-Jensen, U., Bahnfen, M., Dietrichson, O., and Svejgaard, A. 1981. Autoantibodies, histocompatibility antigens and testosterone in males with alcoholic cirrhosis. *Clin. Exp. Immunol.* 44:1, 31-37.
Golub, E. S. 1972. Brain association stem cell antigen: the antigen shared by brain and hemopoietic stem cells. *J. Exp. Med.* 136:369-374.
Graham-Pole, J., Ferguson, A., Gibson, A. A. M., and Stephenson, J. B. P. 1975. Familial dysequilibrium-diplegia with T-lymphocyte deficiency. *Arch. Dis. Childh.* 50:927-932.
Hagberg, B., Hansson, O., Liden, S., and Nilsson, K. 1970. Familial ataxic diplegia with deficient cellular immunity. *Acta Paediatr. Scand.* 59:545-550.
Hurst, E. W. 1932. The effects of the injection of normal brain emulsion into rabbits with special reference to the aetiology of the paralytic accidents of antirabic treatment. *J. Hygiene* 32:33-44.
Kolata, G. 1983. Math genius may have hormonal basis. *Science* 222:1312.
Menzies, C. B., Gunar, M., Thomas, D. G. T., and Behan, P. O. 1980. Impaired thymus-derived lymphocyte function in patients with malignant brain tumor. *Clin. Neurol. Neurosurg.* 82:157-168.
Oldfield, R. C. 1971. The assessment and analysis of handedness: the Edinburgh inventory. *Neuropsychologia* 9:97-113.
Porac, C., and Coren, S. 1981. *Lateral Preferences and Human Behaviour.* New York: Springer-Verlag.
Refsum, S. 1975. Genetic aspects of migraine. In P. J. Vinken and G. W. Bruyn, eds., *Handbook of Clinical Neurology.* Amsterdam: North Holland Publishing Co., vol. 5, pp. 258-269.
Reif, A. E., and Allen, J. M. 1966. Mouse nervous tissue iso-antigen. *Nature (Lond).* 209:523.
Renoux, G., Bizière, K., Renoux, M., and Guillaumin, J. M. 1983. The production of T-cell-inducing factors in mice is controlled by the brain neocortex. *Scand. J. Immunol.* 17:45-50.
Talal, N., Roubinian, J. R., Dauphinee, M. J., Jones, L. A., and Sitteri, P. K. 1981. Effects of sex hormones on spontaneous autoimmune disease in NZB/NZW hybrid mice. In J. Hadden, ed., *Advances in Immunopharmacology.* New York: Pergamon Press, p. 127.

Thomas, D. G. T., Lannigan, C. B., and Behan, P. O. 1975. Impaired cell-mediated immunity in human brain tumours. *Lancet*, i, 1389-90.
Tisserand. 1949. Résultats d'études statistiques sur les becs-de-lièvre. *Semaine d'hôp. Paris* 25:2547-50.
Wikstrand, C. J., and Bigner, D. D. 1982. Expression of human fetal brain antigens by human tumours of neuroectodermal origin as defined by monoclonal antibodies. *Cancer Res.* 42:267-275.

Contributors

Peter O. Behan, M.D. Institute of Neurological Sciences, University of Glasgow (Southern General Hospital, Glasgow, Scotland)*

Charles E. Boklage, Ph.D. Genetics Program, East Carolina University School of Medicine, Greenville, North Carolina

Victor H. Denenberg, Ph.D. Department of Biobehavioral Sciences, University of Connecticut, Storrs, Connecticut

Marian Cleeves Diamond, Ph.D. Department of Physiology and Anatomy, University of California, Berkeley, California

Frank H. Duffy, M.D. Department of Neurology, Harvard Medical School, Boston, Massachusetts (Children's Hospital Medical Center, Boston, Massachusetts)*

Albert M. Galaburda, M.D. Department of Neurology, Harvard Medical School, Boston, Massachusetts (Beth Israel Hospital, Boston, Massachusetts)*

Ida Gerendai, M.D. 2nd Department of Anatomy, Semmelweiss University Medical School, Budapest, Hungary

Norman Geschwind, M.D. Department of Neurology, Harvard Medical School, Boston, Massachusetts (Beth Israel Hospital, Boston, Massachusetts)*

Stanley D. Glick, M.D., Ph.D. Department of Pharmacology and Toxicology, Albany Medical College, Albany, New York

Patricia S. Goldman-Rakic, Ph.D. Section of Neuroanatomy, Yale University School of Medicine, New Haven, Connecticut

Thomas L. Kemper, M.D. Departments of Neurology and Anatomy, Boston University School of Medicine, Boston, Massachusetts (Boston City Hospital, Boston, Massachusetts)*

Marjorie LeMay, M.D. Department of Radiology, Harvard Medical School, and Brigham and Women's Hospital, Boston, Massachusetts (Harvard University Health Service, Cambridge, Massachusetts)*

Gloria B. McAnulty, Ph.D. Department of Neurology, Harvard Medical School, Boston, Massachusetts (Children's Hospital, Boston, Massachusetts)*

Fernando Nottebohm, Ph.D. Field Research Center for Ecology and Ethology, Rockefeller University, Millbrook, New York

Pasko Rakic, M.D. Section of Neuroanatomy, Yale University School of Medicine, New Haven, Connecticut

Steven C. Schachter, M.D. Department of Neurology, Harvard Medical School, Boston, Massachusetts (Beth Israel Hospital, Boston, Massachusetts)*

Arnold B. Scheibel, M.D. Departments of Anatomy and Psychiatry, UCLA Medical School, Los Angeles, California

Raymond M. Shapiro Department of Pharmacology, Mount Sinai School of Medicine, New York, New York

* Mailing address is given in parentheses if different from institutional address.

Index

Aging: and cortical asymmetries, 135–138, 140–143; and hippocampal asymmetries, 143–144
Allergies: and stuttering, 213; and migraine, 219; and mathematical ability, 221
Amphetamine: and circling behavior, 148, 156; and attention deficit disorder, 149; and schizophrenia, 157
Anatomical asymmetry: early research, 1–3; gross, 3, 11–15; radiological visualization, 4, 5, 26–38; developmental, 4, 20–22, 36–37; cytoarchitectonic, 15–20; vascular, 29–32, 76, 86–87; in fossils, 38–40; in dyslexia, 75–88. See also under specific structures
Androgens, see Testosterone
Aphasia: early research, 1–2; crossed, 3; and planum asymmetries, 15; Wernicke's, 16; Broca's, 18
Apomorphine: and circling behavior, 148, 158
Architectonics, see Cytoarchitectonics
Arteriography, asymmetries in: 5, 29–31
Asymmetry: of choline acetyltransferase, 20, 159–160; of norepinephrine, 20, 169; of dopamine, 147–148, 151, 157–160, 174; of glucose uptake, 154–155, 160, 168; of gamma-aminobutyric acid, 159; of glutamic acid decarboxylase, 159. See also Anatomical asymmetry; Behavioral asymmetry; Fossil asymmetry; Functional asymmetry; Pharmacological asymmetry; Radiological asymmetry

Autism, childhood, 213
Autoimmune disease, see Immune disorders
Aversion, learned, 126–127

Behavioral asymmetry: and functional asymmetry, 114–115; and structural asymmetry, 115; measurement, 115–117; models, 117–121; effects of early experience, 122–128; and postural asymmetry, 128–131, 155; hormonal effects on, 129–131
Biological asymmetry: and schizophrenia, 196–197, 199, 201; development, 198–199, 200–201, 202–208; and Down syndrome, 201–202; and fertility factors, 202; in dentition, 204–206
Bird song, see Song learning
Blood flow: radiological imaging, 29–32; and functional state, 55; in dyslexia, 62, 76, 86–87; in rhythm discrimination, 64–65
Brain electrical activity mapping (BEAM), 5, 54; development, 55–59;

Brain electrical activity mapping
(BEAM) *(continued)*
topographic mapping process,
57–59, 67–69; significance
probability mapping, 59, 60; in
dyslexia, 59–63; of responses to
music and speech, 63–64; of
localization of rhythm discrimination, 65–67; digital computer in, 67–68; application, 68–69
Brain tumors, 221
Broca's aphasia, 18
Broca's area: cytoarchitectonic
asymmetries, 19; development, 21;
dendritic structure, 44; in speech
encoding and decoding, 97
Brodmann area 22, 20

Cape of Broca, 14, 18
CAT, *see* Choline acetyltransferase
Caudate nucleus, 160
Celiac disease, 213
Cerebellum, 37
Cerebral dominance, *see* Dominance
Choline acetyltransferase, 20, 159–160
Choroid plexus, 37
Circling: nigrostriatal asymmetry and,
147–148; and learning ability, 149;
and seizures, 149; and cortical
thickness, 150–151; sex differences,
151; inheritance, 152–154; and
reward mechanisms, 155–157; and
dopamine binding, 158–159; and
globus pallidus asymmetry, 160–161
Cocaine, as dopamine inhibitor, 151
Computerized tomography (CT): of
ventricular asymmetries, 27–29; of
hemispheric asymmetries, 31–38; vs.
NMR, 53–54
Convolutional development: normal,
21, 180–183; experimentally altered,
183–186; possible mechanisms,
186–190
Coronal suture, 35–36
Corpus callosum, 127–128, 187–188
Corpus striatum, *see* Nigrostriatal
system
Cortex: Wernicke's area, 3, 19, 43, 77;
auditory association, 16; cytoarchitectonics, 16–19; Broca's area, 19,
21, 44, 97; Brodmann area 22, 20;
laminar depth, 43; in dyslexia,
77–81, 83, 84–87; lesions, 86, 122–128; age differences, 135–138,
140–143; sex differences, 135–140,
151; environmental differences,
140–143; lateral visual, 143; and
sidedness, 150–151; normal
development, 180–183; convolutions, 182–183; experimentally
altered, 183–186; primary visual,
184–186, 188; possible mechanisms
in development, 186–190
Cytoarchitectonics, 15–20, 180

Decoding, 96–97
Delta waves, 67
Dendrites: and human speech, 43–51;
and bird song, 99–100
Deoxyglucose uptake, 154–155, 160,
168
Development: of anatomical asymmetry, 4, 20–22, 36–37; of convolutions, 21, 180–190; of dendrite
systems, 51; neuronal migration, 85,
86, 102–103, 182, 188, 220;
hormones, 99–100, 129–132,
152–153; of functional asymmetry,
154–155; in twinning, 200–201,
203–208; of immune disorders,
220–221
Developmental disorders: and testosterone, 130–131; and handedness,
212–214, 215. *See also* Dyslexia
Differentiation, 40
Dihydrotestosterone propionate,
129–130
Directionality, *see* Spatial preference
Dominance: early research, 1–6; in
dyslexia, 5, 75; anomalous, 5,
214–215; techniques of studying,
6–7; in song learning, 95–96, 98; reversal, 96, 98; for speech, 103–104;
complementary specialization
theory, 103–105; underuse theory,
104–105, 108–109; measurement,
115–117; age differences, 135–138,
140–144; sex differences, 135–140,
143–144, 151; maturational
gradient, 155, 171; and handedness,
159, 214; and endocrine control,
167–176
Dopamine: and circling behavior,
147–148, 151; in schizophrenia,

157; striatal receptors, 157–159; in globus pallidus, 159–160; and prolactin, 174
Down syndrome, 201–202
Dyslexia: and anomalous dominance, 5, 75; incidence, 59; neurophysiological studies, 59–63; and attentional deficit disorder, 62; and handedness, 75, 76, 82, 88, 213, 217; family history, 75, 84, 88, 213; and vascular malformations, 76, 86–87; and migraine, 76, 88, 213, 217; and dichotic listening tests, 77; and seizures, 77, 86–87; ectopic neurons, 79, 85; focal dysplasias, 79–81, 85; verrucous dysplasia, 84–85, 88; hormones and, 88; and immune disorders, 88, 213, 221; and congenital anomalies, 219
Dysplasias: focal, 79–81, 85; verrucous, 84–85, 88; neuronal, 219

Electroencephalogram (EEG): visual inspection, 55–56; computerized analysis, 57, 67–69; in dyslexia, 60–61; in responses to music and speech, 63–64. *See also* Brain electrical activity mapping
Emotionality, 122–127
Encoding, 96–97
Endocrine system: gonadal regulatory mechanisms, 168–171; thyroid control mechanisms, 171–172; prolactin secretion, 172–173; control of grooming behavior, 173–174. *See also* Hormones
Environmental conditions, and cortical asymmetry, 140–143
Ethnic groups, 33–35
Evoked potentials (EP), 55; computerized analysis, 57; visual inspection, 57; in dyslexia, 61
Evolution, 40
Exploration, open field, 122–126

Fear, learned, 126–127
Fertility: and sidedness, 149–150; and twinning, 202
Fertilization response, 203–204
Focal dysplasias, 79–81, 85
Follicle-stimulating hormone, *see* FSH

Forebrain: in song learning, 97; neurogenesis in, 102–103
Fossil asymmetries, 38–40
FSH, 168–169
Functional asymmetry: and dendritic structure, 43–51; neuroimaging techniques, 53–59; in dyslexia, 59–63; in responses to music and speech, 63–64; in rhythm discrimination, 64–67; in bird songs, 93–103; and memory space, 103–105; metabolic activity and, 154–155; endocrine control, 167–174; biological basis, 196–206

GABA, 159
GAD, 159
Gamma-aminobutyric acid, *see* GABA
Genetics: of cerebral dominance, 140–141; of rotational bias, 152–154; of schizophrenia, 196–197; of handedness, 198–199, 214; of twinning, 199, 201–207
Giant pyramidal cells of Betz, 45
Globus pallidus, 157, 159–161
Glucose uptake, 154–155, 160, 168
Glutamic acid decarboxylase, *see* GAD
Gonadotropin, 171
Gonads: and cortical asymmetry, 138–140, 149–150; and hypothalamus, 150, 168–171
Grooming behavior, 173–174
Gyrus: transverse, 13; posterior superior temporal, 16; inferior frontal, 18; angular, 18, 75; development, 21, 180–190; Heschl's, 43; precentral, 44–46

Habenular nucleus, 19
Hallucinations, 157
Haloperidol, 149
Handedness: and language, 2, 3, 46; biological foundations, 6, 198–200, 214; and planum asymmetries, 15; and CT scan asymmetries, 33; and skull shape, 37–38; and dyslexia, 75, 76, 82, 88, 213, 217; and developmental disorders, 130–131, 212–213, 215; and immune disorders, 130–131, 213–214, 217–218; and postural asymmetry, 155; and cerebral dominance, 159,

Handedness *(continued)*
  214; and schizophrenia, 196–197, 199, 201; and twinning, 196–201; and harelip, 212; assessment of, 214–215, 216; and mathematical ability, 221
Handling, 122, 123
Harelip, 212
Hemispheric activation, 116; models, 119; open-field measures, 123; in aversion testing, 127
Hemispheric asymmetries, CT visualization, 31–38
Heredity, *see* Genetic inheritance
Hippocampus: age differences, 143–144; sex differences, 143–144, 151; self-stimulation, 156–157
Hormones: and dyslexia, 88; and song learning, 95, 99–101; and postural asymmetry, 129–131; and immune disorders, 130–131, 220; and cortical asymmetry, 138–140; ovarian, 139–140; and rotational bias, 152–153; and mathematical ability, 221. *See also* Endocrine system
HVC: in song encoding and decoding, 97; destruction of left vs. right, 98; and song repertoire, 98–99; and hormones, 99; growth, 100; seasonal fluctuations, 100–101; neurogenesis, 101–102
Hyperkinetic syndrome, 149, 213
Hyperstriatum ventrale, *see* HVC
Hypoglossal nerve, 95–96
Hypothalamo-hypophyseal-thyroid axis, 171
Hypothalamus: and gonadal regulation, 150, 168–171; reinforcing stimulation, 155–156; and thyroid control, 172

Immune disorders: and dyslexia, 88, 213, 221; and testosterone, 130–131, 220; family history, 214; and handedness, 214, 215, 217–218; and neuronal dysplasia, 219; intrauterine influences, 220; of T-cell function, 220–221
Inferior parietal lobule, 17–18
Inheritance, *see* Genetics
Inhibition, interhemispheric, 116–117; models, 118–119

Lambdoidal suture, 36
Language: and handedness, 2, 3, 46; and dendritic structure, 43–51; EEG responses, 63–64; encoding and decoding, 97; and memory space, 103–104; sex differences, 138
Lateral geniculate nucleus, 184
Lateralis posterior nucleus: cytoarchitectonic asymmetries, 18–19; in dyslexia, 81–82, 85–86
Laterality, *see* Dominance
Learning: neurogenesis, 101–103, 107; lateralization of function, 103–105; network space and, 105; synaptogenesis, 105–107; of aversive response, 126–127; and sidedness, 149. *See also* Song learning
Learning disabilities: and testosterone, 130–131; and handedness, 212–214, 215. *See also* Dyslexia
Lesions: in dyslexia, 76–86; of neocortex, 86, 122–128; left- vs. right-sided, 122–128, 174–175; and mortality rate, 174–175; frontal vs. occipital, 186
LHRH, lateralized control, 168–169
Listening tests: and dyslexia, 77; and inhibition, 117
Locus coeruleus, 169
LSD, 20, 157
Luteinizing hormone-releasing hormone, *see* LHRH
Lymphoid tissue, 219–220

Mastectomy, lateralized effects, 172–174
Mathematical ability, 221
Maturational gradients, 155, 171
Medial frontal lobes, in dyslexia, 62
Medial geniculate nuclei, in dyslexia, 81–82, 85–86
Medulla, 19–20
Memory: and neuronal replacement, 102, 107; and laterality, 103–105; and network space, 105; and synaptogenesis, 105–107; length of, 107–108
Metabolic asymmetry, *see* Biochemical asymmetry
Micropolygyria, *see* Polymicrogyria
Migraine, 76, 88, 213, 217, 218–219

Migration, neuronal: and polymicrogyria, 85; and thalamic malformations, 86; from forebrain, 102-103; and convolutional patterns, 182, 188; intrauterine influences, 220
Minimal brain dysfunction, 149
Morphine, and circling behavior, 156
Mortality rate, and laterality of lesion, 174-175
Music, EEG responses, 63-64
Myeloarchitectonics, 16

Neocortex, see Cortex
Neural networks: size, 98-100, 105; growth, 100; seasonal fluctuations, 100-101; new neurons, 101-103, 107; laterality, 103-105
Neurogenesis: in HVC, 101-102; and memory, 102, 107; in forebrain, 102-103
Neuroimaging techniques, 53-55
Neuronal dysplasia, 219
Neurons: ectopic, 5, 79, 85; dendritic structure, 43-51, 99-100; migration, 85, 86, 102-103, 182, 188, 220; replacement, 101-103, 107; genetic modification, 107; in convolutional development, 187-189; and compensatory enlargement, 189
Nigrostriatal system: and circling behavior, 147-148; and seizures, 149; and cortical asymmetry, 150; hypothalamus and, 150; dopamine receptors, 157-159; prolactin, 174
NMR, 53-54
Norepinephrine, 20, 169
Nuclear magnetic resonance, see NMR

Occipital bone, 37
Occipital horns, 29
Ontogenesis, see Development
Open-field activity, 122-123
Operculum: gross asymmetries, 12, 14; cytoarchitectonic asymmetries, 16, 18; radiological asymmetries, 30; pars triangularis, 44; dendritic structure, 44-51
Orchidectomy and cortical thickness, 138-139, 168
Ovarian hypertrophy, compensatory, 169-171

Ovariectomy: and cortical thickness, 139-140; and lateralized gonadal regulation, 168-171

PCP, 157
PEG, cortical area, 18
PET, scan, 5, 53-55
Petalias, 15, 18, 36-37
PG, cortical area, 17
Pharmacological asymmetry: with LSD, 20, 157; with amphetamine, 148, 149, 156, 157; with morphine, 148, 156, 158; with haloperidol, 149; with cocaine, 151; with PCP, 157
Phencyclidine, see PCP
Pigment architectonics, 18
Pineal calcification, 31
Planum temporale: gross asymmetries, 12-13, 15; cytoarchitectonic asymmetries, 16-18; development of asymmetries, 21
Pneumoencephalography, 26-29
Polymicrogyria, 77-81, 84-85, 188
Positron emission tomography, see PET
Postural asymmetry, 128-131, 155
Prolactin, 172-174
Puberty, 170-171
Pulvinar, 18-19, 20
Putamen, 160
Pyramidal decussations, 19

RA, nucleus: in song encoding and decoding, 97; and song repertoire, 98-99; and hormones, 99-100; growth, 100; seasonal fluctuations, 100-101; neurogenesis, 101
Radiological asymmetry: ventricular, 26-29; vascular, 29-30; by computerized tomography, 31-38
RCBF, 53, 55; in dyslexia, 62; for rhythm discrimination, 64-65
Reading disability, see Dyslexia
Regional cerebral blood flow, see RCBF
Reinforcement, 155-157
Rhythm discrimination, 64-67
Robustus archistriatalis, nucleus, see RA, nucleus
Rotational bias, see Circling

Sagittal sinus, 31, 39
Schizophrenia: and reward processes,

Schizophrenia *(continued)*
157; and handedness, 196-197, 199, 201
Seashore Rhythm Test, 64-67
Seizures: and dyslexia, 77, 86-87; and nigrostriatal dysfunction, 149
Self-stimulation, 155-157
Sex differences: in postural asymmetry, 128-131; in cortical asymmetry, 135-140, 151; in hippocampal asymmetry, 143-144, 151; in sidedness, 151; in dopamine binding, 158; in gonadotropin control, 171
Sidedness, *see* Handedness; Spatial preference
Significance probability mapping, 59, 60
Sinuses, vascular, 31, 39
Skeletal malformations, 216, 217, 219
Song learning: purpose, 93-94; stages 94; deafness and, 94, 98; temporal restrictions, 95; left hypoglossal dominance, 95-96; reversal of hemispheric dominance for, 96, 98; neural encoding and decoding, 96-97; production vs. perception, 98; brain space for, 98-100; growth of networks for, 100; and seasonal fluctuations, 100-101; and neurogenesis in adulthood, 101-103
Spatial preference: handling and, 126; and postural asymmetry, 128-131, 155; and cortical asymmetry, 137, 150-151; and nigrostriatal asymmetry, 147-148, 149; and learning ability, 149; and fertility, 149-150; sex differences, 151; inheritance, 152-154; and dopamine binding, 158-159; and globus pallidus asymmetry, 160-161
Speech, *see* Language
Sphenoid wing, 35-36
Stuttering, 213, 216-217, 219
Substantia nigra, *see* Nigrostriatal system
Sulcus: development, 21, 180-190; central, 36-37
Sylvian fissures: gross asymmetries, 12; architectonic asymmetries, 17; development, 21; radiological asymmetries, 30; in fossils, 38
Synaptogenesis, 105-107

Taste aversion, 126-127
T-cell function, 220-221
Teeth, 204-206
Temporal horns, 26-27
Testosterone: and song learning, 99-100, 101; and postural asymmetry, 129-131; and immune disorders, 130-131, 220; and cortical asymmetry, 138-139; and rotational bias, 152-153
Thalamus: cytoarchitectonic asymmetries, 18-19; chemical asymmetry, 20; CT scan of asymmetries, 37; in dyslexia, 81-82, 85-86; innervation, 187-188
Thymus, 219-220
Thyroid: control of secretions, 171-172; immune disorders, and handedness, 216-217, 221
Thyrotropin-releasing hormone, *see* TRH
Topographic mapping, *see* Brain electrical activity mapping
Tourette syndrome, 213
Tpt area: architectonic asymmetries, 16, 17, 18; chemical asymmetries, 20; in dyslexia, 77
Transverse sinuses, 31, 39
Twinning: and schizophrenia, 196-197, 199, 201; and handedness, 196-201; development, 200-201, 203-208; Down syndrome and, 201-202; fertility factors, 202; polar body, 202-204, 207; tooth development, 204-206

Vagotomy, 169-170, 174
Vascular asymmetries: radiological imaging, 29-32; and dyslexia, 76, 86-87
Ventricular asymmetries, 26-29
Ventrobasal complex, 20
Verrucous dysplasia, 84-85, 88

Wernicke's aphasia, 16
Wernicke's area: gross anatomical asymmetries, 3; cytoarchitectonic asymmetries, 19; dendritic structure, 43; in dyslexia, 77
Word blindness, 75. *See also* Dyslexia